as (honestly!) Mr. Wood. The boys were asked wha
to make, and chose the usual items, such as lette
tstands. I elected to make a boat, so I was given
d and told to hollow out the hull with a hammer a
end of term all the other boys had finished their
but I had not even completed the hollowing-out pr
t the chisel through the bot[illegible]s tha
project was abandoned. Car[illegible]uins.
you will gather, manual dext[illegible]y was not my fort
dent from the outset. I do not actually qualify
xic, but I have to admit that I am fairly close to
-line; for one birthday I was given a tool set, ar
stly ten minutes to hammer my thumb. (Things are r
Quite recently I was trying to fix a tin-opener
wall when someone told me, kindly, that I was pu
de-down. I was.)

t one term at prep school proved to be a false da
me clear that Eton was 'out'. I passed the usual
with the help of tutors, and was geared for Cambr
started. I manœuvred my way into the Forces - I
hat with regard to my age and physique I was deci
cal with the truth - and that put paid to Univers

Eighty Not Out

Eighty Not Out

Patrick Moore

First published 2003 by Contender Books
Contender Books is a division of
The Contender Entertainment Group
48 Margaret Street
London
W1W 8SE
www.contendergroup.com/books

ISBN 1 84357 048 3

Typeset by E-Type, Liverpool
Printed in the UK by Butler & Tanner Ltd, Frome and London
Cover design by Burville-Riley

Endpapers from the original typescript written by Patrick Moore on his 1908 Woodstock typewriter. The text of the printed book follows the style of the original typescript as closely as possible.

Contents

1 The Kid

On several occasions I have been asked to write an autobiography. I can't imagine why. I am not a teenage footballer, a pop star or a rock legend; I am an ancient astronomer. If the total sales of this book amount to fourteen copies, I will not be in the least surprised. However …

I am going to gloss over my first years very briefly. I was born on 4 March, 1923, so far as I know. My birth, unlike Glendower's was not accompanied by any celestial manifestations. My father, Charles – more properly Captain Charles Trachsel Cladwell-Moore, M. C. – was essentially a solder; he died in 1947. My mother, Gertrude – née White – was trained as a singer; of her, more anon. Brothers and sisters had I none.

I grew up first in Bognor Regis and then in East Grinstead, both in Sussex. I was destined for Eton and Cambridge, but never made either, because a faulty heart laid me low for much of the time between the ages of six and sixteen, and there were long spells when I could do very little except read. However, when I was eight I did manage a full term at prep school. I enjoyed it, but not all of the activities were successful. My mother was a talented artist, but this was something which I did not inherit. We had art once a week, taken by a Mr. Moore. On one occasion I was given some art prep, and was told to draw a towel hanging over a chair. I misheard, and thought that I had been told to draw a cow hanging over a chair. I did so. My mother kept that drawing for years; I wish I knew where it is now. Mr. Moore then wrote, saying that I was commendably keen, but on the whole wouldn't it be better if during art lessons I went and played the piano in the music-room? My art career ended at that point.

Carpentry was no better. It was taken by a master whose name was (honestly) Mr. Wood. The boys were asked what they wanted to make, and chose the usual items, such as letter racks and hatstands. I elected to make a boat, so I was given a chunk of wood and told to hollow out the hull with a hammer and chisel. By the end of the term all the other boys had finished their letter racks, but I had not even completed the hollowing-out process, and had put the chisel through the bottom so many times that the entire project was abandoned. Career number two lay in ruins.

As you will gather, manual dexterity was not my forte, as was evident from the outset. I do not actually qualify as being dyspraxic, but I have to admit that I am fairly close to the borderline; for one birthday I was given a tool set, and it took me exactly ten minutes to hammer my thumb. (Things are no better today. Quite recently I was trying to fix a tin-opener to the kitchen wall when someone told me, kindly, that I was putting it in upside-down. I was).

That one term at prep school proved to be a false dawn, and it became clear that Eton was 'out'. I passed the usual school exams, with the help of tutors, and was geared for Cambridge when the war started. I manoeuvred my way into the forces – I have to admit that with regard to my age (sixteen) and physique I was decidedly economical with the truth – and that put paid to university. The other day I happened upon a photograph of a very young Patrick Moore in the uniform of an RAF officer. Looking at it now, I can understand why in those far-off days nobody ever called me anything but 'the Kid'.

What else? – Well, I did have a rather interesting war, but it was long ago, and a great deal of water has passed under the bridge since then. At the end of it I still had my Cambridge option, but it would have meant taking a Government grant, and this did not appeal to me. I prefer to stand on my own feet (size thirteen), so I meant to take up my place as soon as I could afford it. In the event, I never had time, because astronomy took over. I became 'hooked' at the age of six, simply by reading a book, *The Story of the Solar System* by G. F. Chambers, published in 1898 at the exorbitant price of sixpence. I was lucky; a family friend proposed me for membership of the British Astronomical Association when I was eleven, and I was duly elected. (Exactly fifty years later, I became President). In 1952 I was invited to write a book, and I did so: *Guide to the Moon*. It was published a year later, and so this seems to be a rather good place to begin my narrative...

2 The Reluctant Teacher

There are two reasons for beginning these notes in 1953. One, as I have said, is because it was the year in which my first book came out. The second is that this was the year when I definitely abandoned all ideas about taking a conventional job. I was determined to go my own way.

At the end of the war I had no official qualifications – apart from the ability to fly and navigate a turboprop plane – and I had no financial backing at all. My father died in 1947. He and I were quite different people; had he been able to stay in the Army instead of being forced to retire because of a lungful of German gas swallowed in 1917, he would undoubtedly have ended up as a general, whereas nobody could be less military than I am. I was exceptionally close to my mother, who was with me until the day she died: 7 January 1981. One thing was certain: marriage was 'out'. Please do not misunderstand me. I was a perfectly normal boy, I became a perfectly normal young man, and today I am a perfectly normal old man. But Lorna, the only girl for me, was no longer around, thanks to the activities of the late unlamented Herr Hitler; in fact she hadn't been around since 1943. Quite recently, someone asked me whether she was ever in my mind. I replied that after sixty years there were still rare occasions when I could go for a whole half-hour without thinking about her – but not often. This explains why I am a reluctant bachelor, and also why I know that if I saw the entire German nation sinking into the sea, I could be relied upon to help push it down.

When Hitler, the Wops, the Nips and the Vichy Frogs had been disposed of, I had to take some personal decisions. Suddenly, instead of having a great deal of responsibility, I had none at all; a curtain had been dropped, and I was still in my early twenties. My Cambridge place was still there, but I could not overcome my distaste at applying for a Government grant; perhaps illogically, it went against the grain. I was brashly sure that I could write, and that eventually I would earn enough to pay my way through university. At least I could type – and thereby hangs a tale.

When I was six, my grandfather's 1892 Remington typewriter was found in our loft at East Grinstead, and was passed to me as a plaything. Instead, I taught myself how to use it, and I still have it, in full working order; in my will it is left to the Science Museum, where its twin is. Two years later I acquired a 1908 Woodstock; I believe it cost half a crown. By then my 'bible' was W. H. Pickering's book about the Moon, which had been published long ago and was quite unobtainable. A family friend who had connections with a science library in London managed to borrow a copy, and it was in my possession for a month. I remember thinking: 'If I type it out, I'll have the book I want, I'll be able to type quickly, and I'll be able to spell.' It worked like a charm; after 60,000 words I was touch-typing with no difficulty at all. That typed copy is in my study now.

That Woodstock has served me ever since, and all my books, including this one, have been written on it. Mind you, changing a ribbon is quite a business; you have to get a modern ribbon and wind it on to the old spool, a procedure which has to be done about once a week. I have battled with an electric typewriter and even with a word processor, but with a total lack of success. (Not long ago NASA asked me to write a chapter for a book they were publishing about the Moon. I complied, and was sent a reply: 'Thank you for your chapter. This is exactly right, and will go straight to press; moreover, congratulations – you are the first author to send in his material.' And in ink, at the bottom of the letter, a query: 'What the … ing hell did you type it on?')

Writing it would be, but meantime a job was essential. My mother had a modest income, but I did not have the slightest intention of living on that, so a job it would have to be.

The only thing that seemed possible was teaching, so I applied for a post at a boys' prep school, and was accepted. I was sure in my own mind that I was not cut out to be a schoolmaster, but teaching occupied me for the next few years, first in Woking and then at Holmewood House in Tunbridge Wells. Holmewood was new, and when I arrived the total complement was around a couple of dozen boys, aged 8 to 14, and half a dozen adults. John Collings founded the school, and was headmaster; his wife Mary was there, of course. My fellow members of staff were Jo Oldham, circa sixty years old and Alexander Helm, known to most of his friends as Sandy, but to just a few as 'Elm' – because he was once pompously announced by the old school butler as 'Mr. Elm', and the name stuck. He became, and has remained, a very close friend indeed; I was best man at his wedding to Paddy (sadly, no longer here – cancer claimed her) and am godfather to their daughter Pippa.

Setting up a new school must be a pioneering process, and it was certainly true at Holmewood. We were all very close, and as John Collings was not fit – again, due to war service – there were times when Elm and I found ourselves running the show. Once I even had to do the school accounts. The Army had owned the house and grounds before John bought them, and there was a ha-ha round the boundary – a ha-ha being, as I am sure you know, a ditch with one vertical wall, so that it can't easily be crossed. We had no use for it, so we ordered cartloads of soil and filled it in – we had no shortage of boy volunteers, it was that sort of atmosphere. In the accounts I entered 'Dirt for ha-ha, £40.' The auditor marked it 'Ha-ha to you', and sent it back!

I won't dwell on the school years, because I never intended to make teaching a career, and I would have left Holmewood well before 1953 had not John particularly asked me to stay on until everything was really firmly established. In fact I enjoyed my time as a teacher, and I do genuinely think that the boys knew we were there to help them on their way to their public schools. I am still in touch with many of them – and of course some of them have now retired, which is a shattering thought and makes me feel very antique. Four years ago I was in New Zealand, and went to stay for a week with a former pupil, Robert Crawford, whom I hadn't actually seen for a long time even though we corresponded regularly. I think I half-expected to see a twelve-year-old rather than one of New Zealand's most senior and respected surgeons. Tempus fugit ...

Quite a number of Holmewood episodes remain fresh in my memory, by no means all of which would appeal to the modern Politically Correct fanatics and the crackpot child psychiatrists. For example, there was some mysterious creature, which persisted in digging holes in the cricket pitch. We christened it Hubert, and set out to identify it. We put sand round the hole to see if we could find any traces; next morning there were little fairy footprints in the sand – no boy ever 'came clean', though I always had a shrewd suspicion as to the identity of the culprit (I really must remember to ask him next time we have lunch together). Finally we decided upon drastic action. Around 5 November we packed the largest hole with gunpowder, covered it up, and left a trail of powder in the direction of the cricket pavilion. Watched by a crowd of boys, we lit the trail. Fire made its way along the grass, and then into the hole. It was much more violent than any of us had expected, and it was lucky that we were far enough away to dodge flying debris. When all seemed to be quiet we uncovered the hole, which was now so deep that one of the boys wondered whether it would reach through to Australia.

There was no sign of Hubert. Next morning there were six holes in the cricket pitch, and we admitted defeat; Hubert had won hands down. He, or she, or it, was never identified.

I also remember the swimming pool, which we dug with volunteer labour (again there was no shortage of recruits). In the end we had a good pool, suitably lined, refilled every week with water drawn from the mains. Of course it wasn't used in winter, and at the start of summer term we realized that it was (a) full to the brim with green slime on the top, and (b) the plug was in the plughole, with no cord. What to do? We held a poolside conference. Someone, clearly, had to dive in and get the plug out. Elm, who is a very strong swimmer, was not there for some reason or other, and I firmly opted out, because I can barely swim at all. Bill, aged thirteen, came to the rescue. Watched by an admiring host, he stripped off, held his nose, and jumped in. Nothing happened, and I was just wondering how to break the news to his parents when he reappeared, brandishing the plug before clambering out to a round of applause and squelching off toward the showers. To say that he smelt was the understatement of the century...

We did not have a strong religious atmosphere at Holmewood, but officially the boarders were scheduled to go to the local church each Sunday. Somehow or other this never seemed to happen; either it was too hot, or too cold, or urgent cricket practice was needed, or else whoever was due to take church parade had a headache. Eventually the Vicar volunteered to come and take a service in the school dining hall. He was very anxious to have a lectern and a dove of peace. We rustled up a lectern – a prop from one of the school plays – a dove was beyond us. Then one of the boys produced a plastic pterodactyl, and we set it up. I have to admit that it was an evil-looking beast; it leered at you, and I can still see those eyes even now. The Vicar did not regard it as suitable, and went so far as to suggest that we were not taking him seriously enough. Still, we had done our best.

Cricket was always a major sport at Holmewood (it still is), and on Sundays we occasionally fixed up matches against local teams. Some of the older boys were always keen to play, and they were pretty good (both Jo Oldham and John Collings were excellent batsmen, and they coached well; I was less useful, because I was purely a bowler, with an unorthodox action). One evening, after a match, some prospective parents were being shown round the school. Three or four boys were relaxing on the field, and one of the visitors commented that they looked very peaceful. In fact they looked rather too peaceful, and I knew that some cider had been left in the pavilion. Tactfully the parents were steered clear, and subsequently the boys were ushered to their dormitories, very sleepy but with no obvious signs of hangovers. Of course, they had thought that they were drinking something as innocuous as lemonade – and as one of the boys had made an excellent thirty in the match, it was perhaps understandable.

I do not want to give the impression that there was no discipline. There was plenty; for example there was a definite policy about bullying, which amounted to what we would now call zero tolerance. It was accepted that any boy caught bullying would be rather disinclined to sit down for the next hour or two, and the predictable result was that there was no bullying at all – something which would surprise the modern 'do-gooders' who have done such immense harm to our whole educational system.

I have said that I do not propose to dwell on school days, but I cannot pass on without referring to that well-known poet L. F. Antyne. At one stage the boys had been reading poems by T. S. Eliot and came across the immortal lines:

The sunlight shines on Mrs. Porter,
And on her daughter,
They wash their feet in soda-water.

We decided that if Eliot could get away with this sort of verse, so could we, and we invented Antyne. Jo Oldham was superb at this sort of thing, and Antyne poetry became all the rage; it spread like wildfire, and one could not open a boy's English book without finding a new Antyne poem. One which became popular was entitled 'Futility'. I think I wrote it; it may have been a combined effort – anyway, here it is:

The deep futility of all ephemeral things
Which stir the soul to unimagined dreams
Of Brussels sprouts, and spinach in the snow.
The birds' shrill call in the translucent dawn
To embryonic beetles, and pale moths
Which hide their heads in shallow troughs of earth,
Naked and fearful, as the world awakes
To thought transcendent life, and cosmic death.
The earthworm, crawling to his nameless tomb
All energy dispersed, to form new creeds
New auras of the spirits of the wild,
In the deep pool of life, which ceaseless flows
Through endless time and space, in rhythmic praise
Of all creative impulses, which dwarf
The puny concepts of the human mind.
All, all, shall pass into oblivion ...

How sad!

A difficult situation arose when a modern poet came down to give a talk. One wretched boy stood up, read out some Antyne poetry, and asked for comments on it. The poet proceeded to tell us what it all meant. Keeping a straight face was no easy matter, but we managed it, and I must admit that the boys played their parts well; not one of them giggled.

As 1953 drew on, I had to make another decision. I had had my first book published, and interesting things seemed to lie ahead of me – Cambridge or no Cambridge? Things at Holmewood were stable, there was quite a staff, and the numbers of boys were increasing all the time. I was no longer needed, as, frankly, I had been in the pioneer days. So, at the end of the winter term I produced the school plays for the last time, made my farewells, and jumped out into what was, for me, a new and uncharted world.

3 The Craters of the Moon

From 1953 onward, astronomy dominated my life, so I think I must backtrack a little to set the scene. Of course it goes back to the time when I read that little book by G. F. Chambers, and I think I tackled it in the right way.

I did some more reading, obtained a simple star map and learned my way around the night sky, which isn't difficult if you put your mind to it; I made a pious resolve to learn one new constellation on every clear night. Next I borrowed a pair of binoculars, and investigated objects such as double stars and star clusters. By the time I was eleven, I had saved up enough money to buy a small telescope, and I had two slices of luck. One was to be elected a member of the British Astronomical Association; I well remember being taken to a meeting in London, at Sion College, and having the strange experience of walking up to be admitted by the President, at that time Sir Harold Spencer Jones, the Astronomer Royal. It never occurred to me that half a century later I would myself occupy the Presidential chair.

The other slice of luck was that I met W. S. Franks, a well-known astronomer who lived in East Grinstead and ran a private observatory owned by F. J. Hanbury, of the firm of Allen and Hanbury. Brockhurst Observatory was within a couple of hundred yards of my home, and was equipped with an excellent telescope – a 6-inch refractor. Franks took me under his wing, and taught me how to make astronomical observations. He was in his eighties, and just about five feet tall; he had a long white beard, and always wore a skull-cap, so that he looked exactly like a gnome. He was a most delightful man, and it came as a very nasty shock when he died suddenly following a road accident; a car knocked him off the bicycle which he rode every day between his home and the Observatory.

To my intense surprise, I was invited to take over and run the Observatory. For a fourteen-year-old this was a great opportunity, though I have to admit that my main duties were limited to showing astronomical objects to the Brockhurst house guests (Hanbury was mainly interested in growing orchids!). I hope that I acquitted myself well, and of course I had full use of the telescope. My first paper to the BAA was presented during this period; it dealt with features on the Moon, and was entitled 'Small Craters in the Mare Crisium', based on my own work. I sent it in, and was notified by the Association's Council that it had been accepted, but I felt bound to explain that I was not exactly elderly. I still have the reply, signed by the then secretary, F. J. Sellers: 'I note that you are only fourteen. I don't see that that is relevant.' I duly gave the paper, though I imagine that some of the members present at the meeting were distinctly surprised.

For obvious reasons my next paper was delayed until after the war – I had other things on my mind – but afterwards, while at Holmewood, I set up a telescope at East Grinstead. It was a 12½-inch reflector, for which I retain great affection and which is now in my garden in Selsey, protected by a run-off shed. I began regular work, and joined the Lunar Section of the BAA. The Moon was always my special interest, and at an early stage I made a discovery which turned out to be much more important than I realized at the time.

As I am sure you know, the Moon is the Earth's satellite, and moves round us at a mean distance of just under a quarter of a million miles, which astronomically is on our doorstep. It takes 27.3 days to complete one orbit, and it spins on its axis in precisely the same time, so that it always keeps the same face turned toward us; from Earth, there is a part of the Moon which we can never see, because it is always turned away from us. There is no mystery about this. When the Earth and the Moon were formed, about four and a half thousand million years ago (I was away at the time), both were viscous, and raised tides in each other. The Earth is eighty-one times as massive as the Moon, and so its tidal pull was very strong. In fact, the pull of gravity tried to keep a 'bulge' in the Moon facing Earthward, and this slowed down the lunar rotation, rather in the manner of a cycle wheel rotating between two brake shoes. Eventually the rotation relative to the Earth (though not relative to the Sun) stopped altogether, and the 'far side' was rendered unobservable, to the intense annoyance of astronomers such as myself. However, for reasons which need not concern us at the moment, there is a slight, slow wobbling to and fro, so that all in all we can examine a total of fifty-nine per cent of the lunar surface, though of course no more than fifty per cent at any one time.

The edges of the Earth-turned face are very foreshortened. The Moon is covered with mountains and craters, together with broad grey plains which are always called 'seas' even though there has never been any water in them (thousands of millions of years ago they were filled with lava). Most craters are circular, but when seen near the limb (the Moon's edge) they are foreshortened into long, narrow ellipses and are by no means easy to map.

My personal programme was to study these regions, and do my best to chart them, using my 12½-inch reflector. This involved making drawings, and one small incident sticks in my mind. I had come in from the observatory, and was making a fair copy of a sketch, using Indian ink. I had a cup of coffee by me, reached out, took what I thought was the cup, and gulped. Have you ever tasted Indian ink? I don't recommend it, and it is also very difficult to clean it away from your teeth.

One evening at the telescope I happened upon a feature which was not on the official maps, and which seemed to be the nearside edge of a 'sea'. It came into view only when ideally placed, as it was on that occasion. I telephoned H. P. Wilkins, Director of the BAA Lunar Section, who went to his telescope (a 15-inch reflector) and obtained confirmation, so that we could alert our best photographers. It turned out that the feature was indeed a 'sea', but an unusual one; a huge ringed structure extending on to the Moon's far side. This became clear only in 1959, when the Russians sent an unmanned space-craft, Lunik 3, on a round trip and secured the first direct views of the hidden regions.

I submitted a paper to the Association, and suggested a name for the feature: Mare Orientale, the Eastern Sea, because it lay on the east limb of the Moon's face as seen from Earth. This is what we call it today, though an official edict in 1967 transposed east and west – so that my *Eastern* Sea is now on the Moon's *Western* limb.

I was also concerned with what I christened TLP, or Transient Lunar Phenomena, which seemed to be due to gases sent out from just below the Moon's surface layer, disturbing dust and producing elusive glows or obscurations. They are very mild, and for a long time their reality was questioned, but they do exist. Mind you, the Moon is a quiet place; the last major craters to be formed date back at least a thousand million years, so that the dinosaurs must have seen the Moon just as we see it today – assuming, of course, that they were interested enough to look.

After Holmewood I found more than enough to occupy me. I had to earn my living, and by the end of 1953 I did have two books to my credit; otherwise I would have been wary of going freelance, though my mother was all in favour of it. *Guide to the Moon* was one of the books; the other was a translation from the French.

I lay no claim to being a linguist. I speak French with a curious Anglo-Flemish accent, and my grammar is not impeccable, but I am fluent enough (when I go to France I take delight in making out that I can't speak French; some of the comments are most revealing). At one stage I could get by in Norwegian, but now I have forgotten every word. Not that it matters, because all Norwegians speak excellent English.

In 1952, a well-known French astronomer, Gérard de Vaucouleurs, wrote a book about Mars, about which he knew a great deal. He did not then speak English (though he learned it later), and asked me to prepare a translation. I did so, and Faber & Faber in London published it. It is very out of date now, because Mars is not the sort of world we believed it to be fifty years ago, and the marvellous Martian canals have been relegated to the realm of myth. Not that de Vaucouleurs believed in artificial canals or little green men, but he did believe – wrongly, as we now know – that the canals had what he termed 'a basis of reality'. Actually they do not exist in any form, and were mere tricks of the eye. It is only too easy to 'see' what you half expect to see.

But it was *Guide to the Moon* which really sparked off my literary career, and this again was due to sheer luck. As a teenager I had joined the British Interplanetary Society, which was then regarded as rather outlandish; one early member, whom I came to know very well indeed, was Arthur Clarke. When the Society became active again, after 1945, I remained a member, and on one occasion – it must have been in 1950 – I gave a talk about the Moon. By some odd chance a report of that lecture found its way into New York press, and was read by Eric Swenson, head of the publishing firm of W. V. Norton, who was looking around for an author capable of writing a book about the Moon. He checked, and found that I was secretary of the BAA Lunar Section, after which he rang the London publishing firm of Eyre and Spottiswoode. The first I heard about this was in the form of a letter from Robin Warren Fisher, of Eyre and Spottiswoode, inviting me to write a full-scale book.

I was taken aback, but I did realize that this was my big chance, so I sat down in my study, put a new ribbon in the Woodstock, took the phone off the hook, and set to work. For the next few weeks I was seldom seen by anybody apart from my mother and our beloved ginger cat Rufus, who at the age of nineteen was still hale and hearty (it was a sad moment when he finally departed, but at least we gave him twenty happy years). I finished the manuscript, sent it in, and waited anxiously for reactions. I was fully prepared to receive a cold note and rejection slip, but not so; the publishers – both in London and New York – were enthusiastic, and the book went straight into production. It turned out to be a success, and was even reprinted before publication, which was decidedly unusual. It ran to eight editions, and is still in print, though now called *Patrick Moore on the Moon* – a misleading title, because I have not been there. As I have often said, it would take a very massive rocket to launch me into space.

I followed up with *Guide to the Planets*, and this also sold well. Meanwhile, there was another angle which I tried to exploit: science fiction.

I could read easily by the time I was five, and before long I devoured Jules Verne, H. G. Wells, and teenage periodicals such as *Modern Boy*, which I remember as being very good even though it would now be regarded as sexist and jingoistic. Why not try my hand? So the Woodstock was put to work, and in the fullness of time I completed my first boys' novel, *The Master of the Moon*. It was fun to write, but I did not really expect to see it in print. I selected a publisher by sticking a pin into a list, and sent the manuscript off to Museum Press. They accepted it by return. I doubt if I have ever had a bigger shock.

A copy of the book lies on the desk beside me, and I have to say that its scientific credibility is somewhat questionable. Listen to the words of Professor Quinn as he and his companions prepare for take off:

> 'Ought we – ought we to lie down or something?' asked Jock, gruffly. 'I mean, are we liable to be chucked about?'

'Lie down by all means,' said the professor. 'In fact I intend to do so myself, but only as a precautionary measure. The actual start should be no more violent than the sudden ascent of an electric lift.' He settled his spectacles firmly on his nose, and peered at the complicated instrument panel in front of him. 'Ready?'

'Ready,' echoed Sorrell.

'Excellent.' Quinn stretched himself out on the floor, motioning the two boys to do the same and raised his arm.

'I shall count three, and then push the starting button. One... two...'

Noel and Jock watched as though hypnotized. For all the emotion he showed, the little professor might have been asking them for a ride on the Brighton scenic railway instead of shooting them into space at seven miles a second. His hand hovered over the panel, steady as a rock.

'... three,' he concluded, and jabbed.

There was a sudden jerk, and the electric light snapped out, plunging the interior of the rocket into inky darkness. Noel felt as though he were falling into a bottomless pit; yet there was no actual sensation of motion – the walls of the rocket pulsated slightly, and a faint hissing sound could be heard, but that was all.

'Most annoying', came Quinn's voice, irritably. 'Something seems to have gone wrong with the light. Has anybody got a torch?'

Well, you can see the general idea, and worse was to follow. On arrival the space-travellers find a hollow Moon with a breathable atmosphere; a shining underground lake populated by chirping, death-dealing spiders; hideous green cavern-dwellers, and even a villainous Russian who has been there for some time and has declared himself Master of the Moon. Fittingly, the climax comes with a battle with death rays. I enjoyed writing it, and the boy readers seemed to approve, but I doubt whether NASA would view my ideas with much enthusiasm.

In my defence, I plead that when I began to write the novel I had no intention of being scientific; I merely wanted to see whether I could tell a story, and apparently I could. Later on I produced around a dozen more boys' novels, though after the first two I did make an effort to keep to known facts. It was then suggested that I might try an adult novel. I did so; when it was finished I put it aside for a week, and then sat down and read it through. Sadly, I held it over the waste-paper basket and released. It was no good, and I knew it, so that nobody else ever read it.

That was one of the only two manuscripts I finished which never saw the light of day. The other was a farce novel, *Ancient Lights*, which I wrote in 1954 or thereabouts. Had I submitted it, I am fairly sure that it would have been published, but I had so much to do that I never got round to it. There would be absolutely no point in trying it now, because it belongs to a past age. There are no steamy sex scenes, no four-letter words, no persecuted ethnic minority groups and no homosexuals, while the nearest approach to violence comes when the old archaeologist receives a sharp poke in the snoot. It would also be deemed Politically Incorrect by today's standards. I still have the only copy in existence!

Most would-be authors find it difficult to break into the publishing world. I did not, but I am the first to admit that Fate was on my side, just as it was when I began my career in television. Not to put too fine a point on it, I had more than my fair share of luck. At least I took advantage of it; had I not done so, I would not now be writing this book.

4 Facing the Cameras

My very first BBC broadcast was made in 1954, in the overseas service. The subject was about Greenwich Observatory, and I was to join the Astronomer Royal. We were ushered into the studio, and just before we went on the air, 'live', the producer said: 'By the way, this is in French. Do you mind?' Mercifully I didn't, but if it had been any other language I would have been beaten. I felt that I ought to have been warned.

Television came later. I have often been asked how I managed to break into it, and the answer is that I made no conscious effort at all; the idea came from the BBC, and in particular from Paul Johnstone, a highly respected producer who was also a scientist (not an astronomer, but an archeologist). So far as I was concerned, the whole chain of events began with flying saucers.

At that time, flying saucers were all the rage. They had first become headline news in 1947, when an amateur pilot named Kenneth Arnold was flying a private plane over the State of Washington, and reported nine round objects, in formation and travelling at high speed, passing within a few miles of him. He said that they were 'flat, like pie pans', and the term flying saucer came from this.

Arnold believed that he had seen space-ships from another planet; he also believed that the Authorities (with a capital A) were dedicated to playing down his story and concealing the truth. Before long there were more reports, some of them sensational. If they were to be credited, the Earth was under close scrutiny, though by whom or for what purpose remained obscure. Neither was it known whether the saucers were friendly, hostile or merely inquisitive. Photographs of them were invariably blurred, and their whole behaviour was erratic.

I soon began to receive inquiries. Most of the reports could be put down to mundane objects such as birds, bats and balloons; aircraft, searchlight beams and the planet Venus were other favourites. In the early days I remember being telephoned at three o'clock in the morning by an agitated lady living in East Grinstead, who announced that she had seen a saucer hovering over her garden in a menacing manner. When I asked how it was moving, she replied that it 'seemed to be flapping', but I could not convince her that she had been watching a bird on the look out for a breakfast worm.

Where there are space-ships, there must presumably be astronauts, and it was inevitable that there would be reports of contact with alien beings. The spark was ignited by George Adamski, who published the famous book *Flying Saucers Have Landed* in 1953. His co-author was Desmond Leslie, but Desmond's section was confined to historical accounts, comments and interpretation, because he had not seen a Saucer himself. The really sensational part of the book was the Adamski account of his meeting with visitors from Venus, which, he told us, took place at 12.30 p.m. on 20 November 1952.

The place was near Parker, in Arizona, actually on the Californian desert. Together with some friends, Adamski saw the saucer; alone, he located the pilot, who was rather like an Earthman but for the fact that his clothing was atypical for a walk in the desert. For instance, he was wearing what seemed to be ski trousers. Also he had shoulder-length hair, beautifully waved. He could not talk English, and he had to communicate in semaphore, but he was able to establish that his home was on Venus. At the end of the interview the Saucer itself came along to join the discussion, and hovered above the ground; inside there were several people. It was in this remarkable craft that the Venusian made his exit, leaving George alone in the desert.

When the book appeared it became an immediate best-seller. I have no idea how many copies were distributed, but it must have been at least a million. George Adamski became world-famous, and produced several more books; on one occasion he was taken for a ride in a Saucer, and soared close to the Moon. I quote:

'As I studied the magnified surface of the Moon upon the screen ... a small animal ran across the area I was observing. I could see that it was four-legged and furry, but its speed prevented me from identifying it.'

Meantime, I had come to know Desmond Leslie, and we remained great friends until, sadly, he died not so long ago (January 2002). The flying saucer craze showed no sign of abating for some years, and support came from most unlikely people. Notably Air Chief Marshal Lord Dowding, who had been in charge of RAF Fighter Command during the Battle of Britain in 1940. I knew him well. He believed not only in flying saucers, but also in fairies and gnomes, so that he was hardly typical of a senior RAF officer. Yet it was Hugh Dowding who saved us from losing the war. During the Battle of Britain he did not make a single mistake, which was just as well in view of the fact that we had no reserve aircraft. It was also he who persuaded Winston Churchill not to send any aircraft to France after the evacuation from Dunkirk. In my view Hugh Dowding was one of the greatest Englishmen of the twentieth century, and he never received full recognition for what he achieved at a time when we needed him most.

A television programme was planned. Desmond, Hugh Dowding and a couple of other Saucerers were to present the case in favour of flying saucers, and Desmond asked me to present the case 'against', because he was well aware of my lack of faith in all forms of airborne crockery. The programme was duly broadcast, and was well received. Paul Johnstone produced it, and said that he was extremely satisfied.

We went our separate ways, and I thought no more about it until one morning when I had an interesting phone message. Would I go to Lime Grove, at that time the headquarters of BBC television, and talk to Paul Johnstone about a possible programme dealing with astronomy?

I was elated – well, who wouldn't have been? – and I lost no time in making my way to Lime Grove. It transpired that Paul had been casting around for an astronomy presenter; he had read a book of mine and also studied the Saucer programme, with the result that he thought I might be suitable.

We had a long talk. What exactly was our aim? Could we manage a programme every month, slanted toward the newcomer and yet with enough solid material to entrap the well informed? Paul thought we could, and so did I, though at the time my knowledge of television was effectively nil, and in any case television had nothing like the influence it has today. There was also the problem that astronomy was widely regarded as being practised only by old men with long white beards, who spent their lives in lonely observatories 'looking at the stars', and no doubt using crystal balls as well.

Since all this happened more than forty years ago, I think I must say a little about the situation at that time. In early 1957 the Space Age had not started; it was not until the following October that the Russians launched Sputnik 1, the first artificial satellite, which took many people by surprise and caused considerable alarm in America. (One man in Washington rang the Pentagon to say that a sputnik had landed in his garden and was lodged in the top of a high tree; it turned out to be a balloon, with 'upski' painted on the top and 'downski' on the bottom.) The great radio telescope at Jodrell Bank was only just being brought into action, and was widely regarded by the general public as being a waste of money. The idea of sending a man to the Moon was little more than a music-hall joke, except to those people who had taken the trouble to investigate, while rockets were still associated only with the V.2 weapons masterminded at Peenemünde by Wernher von Braun. Generally speaking, anyone who could recognize the Pole Star and the Great Bear was doing rather well.

Television at that time was decidedly primitive as well. Of course it was all black and white, and because recordings were not up to standard everything had to be 'live', so that disasters were not uncommon. There was little in the way of electronics, and producers were compelled to use all sorts of dodges, as I was to learn very soon in my BBC life.

One initial problem we faced was the selection of opening music, which is actually more important than might be thought. The usual suggestions, such as 'You Are My Lucky Star', were rejected out of hand. We also dismissed Holst's *Planets*, partly because it was too obvious but mainly because it was astrological anyway. Finally we hit upon a movement from Sibelius's suite *Pelléas et Mélisande*. It was called 'At the Castle Gate', and it proved to be a great success. We still use it, and have not thought of changing it. Only once did we vary; for our last programme about Halley's Comet, in 1986, the Band of the Royal Transport Corps played us out with my own march, appropriately called *Halley's Comet*.

In pursuit of 'props', we went to see Alfred Wurmser, a charming Viennese who lived in Goldhawk Road, he had a dog named Till, half-Alsatian and half-Wolf, who weighed about a ton but was under the strange delusion that he was a lap-dog. Alfred made moving diagrams out of cardboard, and he soon became enthusiastic, so that we continued to use the 'wurmsers' until he decided to return to his native Austria. The original title of our programme was to be *Star Map*, but we changed it to *The Sky at Night* almost at once – to make sure that the new title went into the *Radio Times*.

At that stage we had a stroke of luck. A bright comet appeared, and caused a great deal of popular interest. Could it be an omen?

Comets have often been classed as unlucky, and held responsible for dire events such as plagues, earthquakes and even the end of the world. We know better. A comet is a ghostly thing, and has been described as a dirty iceball, so that it could not possibly knock the Earth out of its orbit; one might as well try to divert a charging hippopotamus by throwing a baked bean at it. The new comet, Arend-Roland (named after the two Belgian astronomers who had discovered it) was unusual inasmuch as it appeared to have two tails, one pointing away from the Sun and the other toward it. Actually the sunward 'tail' was due to nothing more than fine dust spread along the comet's path, but it looked intriguing, and for several evenings during April the comet was easily visible with the naked eye.

Obviously, Arend-Roland had to be the centrepiece of our first programme. I took photographs of it, and managed to obtain others; we revised our original plans, even relegating an eclipse of the Moon to the last few minutes of the programme. Eventually we were ready for transmission – or so we hoped. Remember, I had been on television only once before, and I had no real idea of what to expect, particularly as I had no guest appearing on the programme with me.

Rehearsals seemed to go well. Even at that stage I had no word-for-word script; Paul trusted me to bring in the visuals (photographs, diagrams and 'wurmsers') at the right moments, and otherwise it was up to me. And so at 10.30 p.m. on the evening of 26 April 1957 I was seated in my chair in the Lime Grove studio waiting for the red light over the television camera to come on.

Was I nervous? In a way I suppose I was; I remember thinking 'My entire life depends upon what I do during the next fifteen minutes.' Then the screen on the monitor began to glow; I saw the words '*The Sky at Night.* A regular monthly programme presented by Patrick Moore', and the series was launched. It did not then occur to me that I would still be broadcasting more than forty-five years later.

I waited anxiously for reactions to that first programme. I was grateful to that twin-tailed comet, and I was very sorry to see it depart from our skies a few weeks later. It will never return – it had the misfortune to pass by the giant planet Jupiter, and was unceremoniously hurled out of the Solar System altogether, so that even if *The Sky at Night* is still being broadcast a few tens of thousands of years hence, my successor will be unable to welcome Arend-Roland back. Meanwhile, we all wanted to know whether we had made the programme (a) too elementary, or (b) not elementary enough, or (c) pitched at about the right level.

This is always a problem, and I can only hope that we have guessed right. It is true that the programmes today are watched by many professional astronomers as well as amateurs, no doubt because the field has become so vast that nobody can hope to cover it all: for instance, a student of remote galaxies need not necessarily know much about the polar caps of Mars. We also have to cater for viewers comfortable in their armchairs watching the programme because they have not mustered up enough energy to get up and switch off. Our aim is to capture their interest as well.

Within a day or two of our first foray, letters began to flood in, and most of them were encouraging. I answered them all; as with rare exceptions I still do; the Woodstock works until it is almost red-hot. Mind you, there are occasions when I have been baffled. One early correspondent wrote to me saying that he had enjoyed hearing me talk about comets, and in consequence now wanted to buy an Army tank: had I got any? There was also the dear lady who was anxious to communicate with the beings who live on the Moon, and wondered whether it would be possible to send a carrier pigeon there. I suggested sending a pigrot – i.e. a bird which was a between a carrier pigeon and a parrot, very useful, as it could convey verbal messages.

Flying saucers will not go away, and I remember one episode

in July 1963, when a peculiar crater appeared in a potato field at Charlton – not the London Charlton, but a small village near Shaftesbury, in Dorset. A local farmer made the discovery, and also saw that the crops over a wide area around had been flattened. Reports over the radio and in the press caused widespread interest, and this was heightened by a statement from an Australian who gave his name as Robert J. Randall, from the rocket proving ground at Woomera. Dr Randall maintained that the crater has been produced by the blast-off of a saucer from the planet Uranus. It was independently suggested that there might be a bomb in the crater, and an Army disposal squad was called in.

When the whole affair started to look really interesting, I happened to be in a television studio. We decided that whatever was happening, we ought to be 'in' on it, so at the dead of night we drove to Charlton, arriving in the early hours to find all sorts of people hopping about like agitated sparrows. The bomb disposal squad was at work, but had unearthed nothing except a small piece of metal which might have been anything. As I knew something about dismantling bombs, I was called in, and I had no qualms, if only because I thought that the chances of there being any buried explosives there were about a million to one against. There were also a water diviner, marching around with an impressive assortment of twigs; there were several astrologers, at least one telepath, and various local Saucerers. The teeming populace was kept away by improvised fences, though I was allowed to go where I liked. The crater was evident enough. It looked as though it had been caused by subsidence, but more than that I could not really say.

We then tried to locate Dr Randall, but could find only a relative of his who seemed to be the local district nurse. Strangely, Woomera disclaimed all knowledge of anyone of that name; we went so far as to telephone Australia. So far as I know, nobody has seen him since, though some time later he did issue a report on the Charlton affair, adding that on another occasion he had come across a grounded Uranian saucer and had had a long conversation with the pilot, who rejoiced in the name of Ce-fn-x.

I did track down the rumour that when the spacecraft landed, it killed a cow. What had happened was that during a discussion at a local pub, a farmer had said 'Ah! and, you know, a cow of mine died last week, too!' with the inevitable result that a journalist overheard, and another sensational headline was dispatched to a newspaper.

I drove past Charlton a few weeks later, but all was quiet, and there have been no further cosmic visitations. By then the main interest had switched to Warminster, a pleasant little Wiltshire town. Various sightings had been reported from there, mainly from the adjacent Cradle Hill, and eventually I went to Warminster with a television team, arriving soon after nightfall. It proved to be a fascinating experience, but Saucers failed to put in an appearance.

Only once have I been really thrown. This was in 1977, when I was carrying out some observations of the Moon with my 15-inch telescope. Suddenly a whole crowd of Saucers came toward me. They moved slowly but deliberately; in the telescope field, they glowed with a strange, eerie light, and they were genuinely Saucer-shaped. As I watched, mesmerised, they sheered off and vanished. I simply did not know what they were, and only in the following day did I find the answer: pollen, catching the rays of moonlight and being seen totally out of focus.

The Saucer craze has waned now, even though a few years ago the European Parliament seriously considered setting up a helicopter base to accommodate visiting aliens (an idea absolutely typical of our would-be masters in Brussels). No doubt new landings will be reported at some time in the future.

I was once asked what I would say if a Saucer landed on my front lawn, and a little green man emerged. I replied that I knew exactly what I would say. 'Good afternoon. Tea or coffee? Then do please come with me to the nearest television studio.' There is nothing I would like better than to interview a Martian, a Venusian or even a Saturninan, but somehow I don't think that it is likely to happen.

5 Pioneers

For me, the *Sky at Night* 'trial period' was decidedly nerve-racking. Having started appearing on television I was anxious to go on, but that was up to the BBC; of course there was no commercial TV in those days, and BBC1 was on its own. June 1957 came and went, and there was no suggestion of taking the programme off the air. In fact, so far as I know there never has been.

Why not? I think I know. Of course it is the subject which matters, not the presenter; moreover the *Sky at Night* is nobody's enemy, because it is non-controversial, it is cheap to put on, and it is shown late at night, so that it does not interfere with the endless quizzes, police dramas and kitchen-sink 'sitcoms'. I may add that I have never had any sort of long-term contract with the BBC, and I merely sign what is put in front of me, and in theory I could present astronomical programmes on other networks. Frankly, I have had plenty of offers (with excellent financial terms!) but not for a moment would I consider them. I have a gentleman's agreement with the BBC, and if I am asked to take part in a programme elsewhere I check to make sure that there can be no overlap. A contract can be broken; a gentleman's agreement cannot – at least, not so far as I am concerned. I may be regarded as a dinosaur, but that is the way that I'm made.

I remember one of these very early programmes for an unusual reason. Paul and I had agreed to invite guests, and among the first was one of the greatest astronomers of the twentieth century,

Harlow Shapley, the man who first measured the size of our star system or Galaxy. As I explained in the lead into the programme, the Galaxy, with its hundred thousand million stars, is shaped like a double-convex lens, or, more descriptively, two fried eggs clapped together back to back. (My original wish to produce two eggs, fry them in front of the camera and then make a sandwich out of them, had been vetoed by Paul). Round the Solar System are clumps of stars known as globular clusters, all of which are so far away that their light, travelling at 186,000 miles per second, takes thousands of years to reach us. Harlow had been able to measure the distances of these clusters, by using special stars which 'give away' their real luminosities by the way in which they behave. He was then able to establish that a ray of light would take 100,000 years to speed across the Galaxy from one side to the other. He also found that the Sun, with its family of planets, lies well away from the centre of the system, so that we have a decidedly lop-sided view.

Also on the programme was another great astronomer, Bart Bok. We were sitting in a line – one, two, three – and we had to have what was known as a recording break. We had filmed a short section of the programme earlier, to avoid an awkward and complicated camera move, and this insert was put into the live transmission as it came out. With malice aforethought, and with thoroughly evil intent, Harlow and Bart changed places for the live transmission, so that when we came to the insert they flicked to and fro in a most bewildering manner. By the time that Paul and I found out what they had done, it was too late...

Alas, both Harlow and Bart are dead now. They were splendid people as well as famous astronomers, and they are much missed.

By the late summer of 1957 I was starting to feel that *The Sky at Night* was well established. At that time astronomy was classed as a minority interest, but on 4 October came a new development; the Soviet Union launched Sputnik 1, the first artificial satellite, and ushered in the Space Age. I doubt whether any single event has done more to change the world. In time to come, few people will remember 1066 or 1914, but 1957 will not be forgotten so long as humanity lasts.

Sputnik 1 was not unexpected, because it had been an open secret that both the USA and the USSR has been working feverishly to put a satellite into orbit. When Sputnik soared aloft, some Americans were far from pleased, particularly as their own space programme was in a state of disarray. In general their rockets either blew up, fizzled out, or crash-landed soon after blasting off; few people were allowed to visit the proving ground at White Sands, in New Mexico. Soviet Russia, of course, was completely sealed off from the outside world, and very little information was released. In 1955 I had written a book called *Earth Satellite*, in which I gave the facts as far as I knew them, but inevitably there was a great deal of speculation.

At least I did know some of the main characters in the spaceflight story, and this may be the moment to say something about them, because none of them are left now.

To begin with, there is Orville Wright – yes, the first of all airmen, who made that initial 'hop' from Kitty Hawk more than a century ago now. I met him at the start of the war, when I was learning how to fly; I had about half an hour's conversation with him, and it is something that I will never forget. He was quiet, unassuming and I thought slightly sad, because he made no secret of the fact that he hated the idea of aeroplanes being used to drop bombs on people below. He did little flying after about 1920; his brother Wilbur had died in 1912, and Orville seemed content to retire gracefully into the background. It is rather sobering to reflect that Orville Wright and Neil Armstrong could have met. They never did, but their lives overlapped.

To digress for a few moments, I must add that during the same period I met Albert Einstein, who turned out to be precisely the sort of person I had expected; I could well have believed that he had come from another planet. He was utterly charming to everybody, though of course this was my sole encounter with him. We met at a small reception, and there was one amusing episode. As we all know, Einstein was a talented violinist, and on this occasion he had a violin with him – he had been playing in a private concert. Pressed to show his skill, he said that he needed an accompanist. There was a piano to hand – and so there was Einstein playing Saint-Saens' *Swan* to my accompaniment. O for a tape! But there were no tape recorders in those days, and whether there is any recording of Einstein as a violinist I do not know, if there is, I would like to obtain it.

Most of my meetings with the space pioneers came later, when the hot war was over and the cold war was in full swing. I well remember Hermann Oberth, the Romanian theorist who wrote the first really scientific book about space research, in 1923, the year I was born. I met him on several occasions during conferences in the 1960s, but there were problems, because he spoke no English at all and my German is limited to 'Damit!' 'Besonders!' and 'Donner und blitzen!'. Oberth will always be remembered for his one classic book, though to be candid he did not accomplish much else, and so far as I could make out he stood politically well to the Right of Genghis Khan. Still, I am glad to have met him. He was, incidentally, a man of great personal courage, as he showed during the RAF raid on the Peenemünde rocket base in 1943 – and this brings me on to the key figure of those early years, Wernher von Braun.

The idea of space-travel goes back a long way, but until the development of rockets it remained a wild dream. Aircraft, such as those of the Wrights, or for that matter the Concordes of today, depend upon having air around them, and there is not much air above a few tens of miles, so that aircraft won't work. Rockets, on the other hand, depend upon what Isaac Newton called the 'principle of reaction', every action has an equal and opposite reaction, so that a rocket will, so to speak, push against itself. If this baffles you, blow up a balloon and then suddenly let the air rush out. The balloon will shoot across the room, because it is being pushed by the air streaming out of its exhaust. In a rocket, a jet of gas is produced by a special kind of motor, usually involving a 'fuel' and a 'propellant', which when mixed together generate heat and send the gases out through the rocket exhaust. Obviously, surrounding air is not needed, and is actually a nuisance, because it sets up resistance and has to be pushed out of the way. This is why rockets, and rockets alone, can be used for flight above the top of the atmosphere. The first man to realize this was a Russian, Konstantin Eduardovich Tsiolkovskii, near the end of the nineteenth century.

Tsiolkovskii was purely a theorist, and never fired a rocket in his life. The first liquid propelled rocket was set off in 1926 by an American, Robert Hutchings Goddard. (Solid fuels have marked limitations, because they are not controllable, and have an unpleasant habit of going bang.) But Goddard was the reverse of publicity-minded, and the research was taken up by a group of German enthusiasts, who established a rocket proving ground outside Berlin and began tests. One of these enthusiasts was von Braun, who was very young but also very clever indeed.

There were mistakes and setbacks, but the Germans persevered, and in time began to make good progress. There were some amusing episodes, too – notably the Magdeburg Experiment. Some members of the Magdeburg City Council became convinced that the Earth is the inside of a hollow sphere, so that a rocket launched vertically upward at sufficient speed would land in the Antipodes. They therefore provided funds to send up a rocket for precisely this purpose. Von Braun and his colleagues had no faith in hollow Earths, but they did need cash, and two rockets were actually dispatched. As one of them exploded immediately after take-off and the other travelled horizontally instead of vertically, the results were inconclusive!

On a more sinister level, the German Government realized that rockets could be of immense value in war. The original testing ground was taken over, and the research transferred to Peenemünde, an island in the Baltic. Here, the famous – or infamous – V2 was developed, and used to bombard Southern England during the closing months of fighting.

For a long time the Allies did not know what was going on. Peenemünde was highly suspect, but Lord Cherwell, chief scientific adviser to Mr Churchill, was sceptical, and it was only in 1943 that a bomber attack was made by the RAF. The main idea was to kill the scientists as well as destroy as much equipment as possible. (I am glad that I was not on that raid, though I might well have been). The attack was a partial success, and it was then that Hermann Oberth showed his bravery, saving several lives and pulling people out of burning buildings. Of the scientists, only one was killed – Dr. Thiel, head of the propulsion department, together with his whole family.

The raid did not halt the development of the V2, but it did mean that some departments were transferred underground, safe from attack. Luckily for us, time was against the Germans. They were losing the war; the V2 was developed too late to turn the tide, and eventually it became clear that the game was up. Peenemünde was abandoned, and although a few of the leading scientists went to the USSR most of them, including von Braun, surrendered to the western Allies, and were taken en bloc to the United States. It was not long before von Braun held a vitally important position in American rocket research. I first met him around 1952 – I do not remember the exact date – when it was obvious that artificial satellites would be launched in the near future.

I doubt if anyone could loathe the memories of the Nazis any more than I do. They killed my girl, many of my best friends, and did their best to kill me, and their concentration camps, believe me, were unspeakable. When I go to Germany today (which is not often) I am still conscious that I am in enemy territory. So when I first met von Braun, I expected to recoil and to be reluctant to shake his hand. Yet to my surprise that was not how I felt, and there was no sense of instant dislike. We shook hands, we talked, mainly about the Moon, and we were quite at ease with each other; neither did he react to the RAF aircrew tie which I was wearing, and which I still wear to this day (or rather, another tie to the same design!). I also found that unlike most of his countrymen, he had a sense of humour, and I strongly suspect that he was responsible for the celebrated notice posted in White Sands when the German contingency arrive there:

I met him quite often in later years, and we even broadcast together; there was no hint of friction between us. It was all decidedly strange, and I have often tried to work out the answer.

Make no mistake about it: some of the German rocket scientists were guilty of atrocities, and conditions in the 'Dora' camp, outside Peenemünde, were horrific. Slave workers were brutally ill-treated and often killed, and the situation was not much better than that in a concentration camp. One very senior rocket scientist Arthur Rudolph, was eventually forced to give up his American citizenship; he left the United States to avoid prosecution for war crimes, and returned to Germany. By that time Wernher von Braun had died of cancer. Could he, too, be classed as a war criminal?

Some people think so, mainly, I suspect, because he had joined the Nazi Party. But in this he had no choice. Had he refused to become a member, he would not only have been removed from his position at Peenemünde, but would almost certainly have ended up in prison. In fact Hitler did once have him arrested, believing (correctly) that he was much more interested in the Moon than in building military weapons. He was certainly not an enthusiastic Nazi, and wore his uniform only when there was absolutely no alternative.

The other point at issue is whether he knew what was going on in Dora and the other slave camps, as others – notably Arthur Rudolph – unquestionably did. We will never know for certain, and the jury must remain out, so that all I can do is to give my own opinion, with the full admission that I may be quite wrong. So I am prepared to say that I do not believe von Braun to have been guilty of anything but inevitable lip-service to the Nazis, and I do not believe that he took part in any atrocities. There we must leave it.

(As an aside, von Braun was quite unlike the austere, monocled German of the Erich von Stroheim type. He did not wear a monocle. I do, and I have often been asked why. When I was sixteen I went to an oculist, who found that I would benefit from a lens for one eye, not the other, and suggested a pair of glasses with one plain lens, I saw no sense in that. 'What about a monocle?' He replied, 'Won't a boy with a monocle look rather odd?' I didn't mind, so I acquired a monocle. It is a weak lens, and I can see quite well without it, but it simply means that my two eyes are equal when I wear it. I tend to feel undressed without it!).

In the old USSR, the leading rocket scientist was Sergei Korolev, the 'Chief Designer', who survived Stalin's purges and became the Soviet equivalent of von Braun. He died suddenly after what ought to have been a routine operation; had he lived, the whole story of Russian rocketry might have been different. He and von Braun never met, of course, and during the Cold War there was no contact whatsoever between the two rocket teams. It was only in the dying days of the USSR that the situation changed, and people such as myself were able to go to the Russian rocket centre at Baikonur. I have met many of the cosmonauts – people such as Alexei Leonov and Yuri Gagarin – and they have many points in common with the American spacemen. Beyond Earth, nationalistic prejudices seem to evaporate, and sanity takes over.

6 Here and There

For me, the years between 1957 and my departure from East Grinstead, in 1965, were totally dominated by *The Sky at Night* television programmes, though I had many other interests as well, and I was busy writing both popular astronomy and boys' science fiction. I was a very active member of the British Astronomical Association, and became Director first of the Mercury and Venus Section, devoted to studies of the two inner planets, and then of the Lunar Section. The Moon was my main interest (it still is), and I became the junior partner in the compilation of a large lunar map, drawn mainly by the Welsh amateur Percy Wilkins. By profession Wilkins was a Civil Servant working for the Ministry of Supply, but whenever I visited him in his office I found that his desk was covered with drawings and photographs of the Moon; of course, as a civil servant he had plenty of spare time. He retired in 1960, with the intention of devoting the rest of his life to astronomy. Sadly, it was not to be; he died suddenly a few months later, and was much missed.

By this time the Moon, previously neglected by professional astronomers, had become of paramount importance, and the 'space race' was well under way. At first the Russians took the lead, and launched Sputnik 1; only some time later did the first American satellite, Explorer 1, follow. It could have come earlier if von Braun had been given full backing, but inter-service rivalry in the United States had led to delays, and some of the launches were failures, leading to the famous Cape Canaveral count-down, a bowdlerized version of which runs 10–9–8–7–6–5–4–3–2–1–*bother*! It is also on record that one vehicle landed in Cuba and killed a cow; there was an official protest, and the cow was given a state funeral as a victim of Imperialist aggression. Then, in 1959, the Soviets sent up their first three lunar probes.

Lunik 1 was the first. In January it by-passed the Moon, and sent back useful information, such as the fact that the Moon has no overall magnetic field, so that ordinary compasses will not work there (I must remember that, next time I go for a stroll in the Mare Tranquillitatis). Then, in September, came Lunik 2. It was launched on the twelfth, and was scheduled to crash-land on the eighteenth. By now *The Sky at Night* was fairly well established, and we were able to put on the first of many 'special' programmes, so that on the evening of 12 September we were ready to go.

In a very minor way I had myself been involved in the purely scientific research. As I have said, I was a Moon mapper; the Russians had asked me to send them my results, which I had done, and they were very good about keeping me abreast of the latest developments. A Soviet astronomer happened to be in London, and although I had never met him it seemed a good idea to call him in. Paul agreed. A hasty message was sent, and we were assured that our guest would turn up on schedule.

Naturally we were 'live', but when the programme started there was no sign of the Russian. Unfortunate, I thought, but perhaps he would appear, if not, I could manage without him. After a couple of minutes, when I was well launched into an account of what we hoped Lunik 2 would tell us, I saw a newcomer being ushered in by the floor manager. As they approached, the floor manager held up a message for me to read: 'HE DOESN'T SPEAK ANY ENGLISH'.

What to do? I decided to take a gamble, and asked a question in English, emphasising the word 'Lunik'. He replied, in Russian.

For all I knew, he might have been describing collective farming in Omsk, but I said in English what I hoped he was saying in Russian – and we did the whole programme like that. At the end we went off the air, and then the mystery was solved; we had been communicating by post – but in French, which we both spoke. I waited for indignant letters from viewers who could understand Russian, but I didn't get any, so apparently my 'translation' had been fairly near the mark.

The next Soviet probe of 1959 was Lunik 3, which began its journey on 4 October, exactly two years after the epic flight of Sputnik 1. This time I had all the information I needed, because I was involved more as an astronomer than as a television presenter.

Remember that the edges of the Moon as seen from Earth are very foreshortened, and that forty-one per cent of the lunar surface is permanently turned away from us. The Russians had sent for my maps of the foreshortened areas, because they wanted to use them to link the far side pictures with features on the known side, and they had promised to send me the Lunik 3 images as soon as they became available. They were due on 24 October, by which time the probe had been right round the Moon and had obtained the pictures – which were then scanned by an on-board television camera before being transmitted to Earth.

In Lime Grove, we were ready. Meanwhile, unknown to me, the first pictures had arrived in London; the Russians had kept their promise. A BBC messenger was waiting; he grabbed the pictures, jumped on to his motorcycle, and drove to Lime Grove at a pace which would certainly not have been approved of by the traffic police. Five minutes into the programme, I had a message from Paul Johnstone in the gallery. 'First views of the Moon's far side coming up on the screen in thirty seconds. Scrap what we'd planned. Do it off the cuff!'

It was a tremendous moment, and I knew that I was about to see something that I had wanted to see all my life. I took the audience into my confidence: 'I don't know what's going to come up, but it's bound to be exciting... There it is. Look at that!'

Frankly, it was not eye-catching, because, as I realized, the Lunik pictures had been taken at a range of over 40,000 miles from the Moon's surface, and there were virtually no shadows, because the Sun, the rocket and the Moon were almost lined up, giving the equivalent of full-moon illumination. However, I was able to recognize one feature, the Mare Crisium or Sea of Crises, because it can be seen from Earth (it had been the subject of my first paper to the BAA, a quarter of a century earlier). As soon as I had my bearings, I was able to give what I hope was an intelligible commentary. We had been right in saying that the far side of the Moon was just as cratered, just as rough and just as barren as the familiar side. George Adamski's green fields and little furry animals were conspicuous only by their absence.

The Lunik 3 programme remains one of the main *Sky at Night* highlights so far as I am concerned. The timing was fortuitous; whether we would have been granted a special programme I do not know, because all the networks were waiting to cover the story, and it was sheer luck that we happened to be transmitting at exactly the right time.

I thought that the Russians would follow up their triumph quickly, but for some time there was a lull, and it was April 1963 before they tried again. Lunik 4, alas, was not so successful as its predecessors, as we found out to our cost.

The aim, as the Russians had told me, was to land a capsule on the Moon gently enough to avoid it being damaged. It could then transmit direct from the lunar surface. We mounted a special programme (in those days space travel still took priority over a football match), and made quite elaborate preparations. We had a telephone link with Moscow, a radio link with Sir Bernard Lovell at Jodrell Bank, and cameras mounted on large telescopes at Brighton and at the Royal Observatory, Edinburgh – the idea being to get the latest news from Moscow, listen to the signals from Jodrell Bank and then survey the landing area from Brighton, Edinburgh or both. We had a live transmission lasting for an hour and a half.

Do you know those times when absolutely nothing goes right? It was so on this occasion. Nobody in Moscow knew anything; nobody in Jodrell Bank could hear anything; it was raining in Brighton, it was cloudy in Edinburgh, and moreover Lunik 4 missed the Moon by a full five thousand miles. That was one of my early experiences in what is known in BBC jargon as 'padding'. Still, you can't win all the time, and at least we had broken even, with one and a half successes out of three Moon shots.

It was also around this time, during one of the Moonshot programmes, that I swallowed a large fly, an episode which is remembered even today. We were on the air, live as usual, and when I opened my mouth to make some world-shattering announcement the fly flew straight in. I don't know what size it was – at the time, it felt like a bluebottle – but to my eternal credit I swallowed it, and it buzzed all the way down. Paul told me that he saw a look of glazed horror come into my eyes, after which I gave a strangled gulp and went on. My mother summed it up rather neatly. 'Yes, dear,' she said. 'Nasty for you, but how much worse for the fly!'

Rather to my surprise, I was invited to go to the USSR as a guest of the Academy of Sciences, my rôle being to take part in discussions about lunar landing sites and give a couple of lectures. I was also made an Honorary Member of the USSR Astronomical and Geodetic Society. Whether this is still valid, as the USSR has long since departed, I know not. Anyway, it was a fascinating visit, my first experience of rigid Communism. At least the scientists were as friendly as they could possibly be, and I was made welcome wherever I went. My guide and companion was a student named Gregory Molyikov; I wonder where he is now?

Halfway through the trip, Gregory took me to Moscow station and put me on the train bound for what was then Leningrad, where I was scheduled to address a meeting. The carriage was clean and comfortable (British railways, please copy!) and was occupied by one Russian, who was smoking a pipe and gave me a welcoming nod. Clearly he was happy to talk, but there were problems. We tried English and French: no good. Suddenly he pulled out a travelling chess set, and invited me to have a game. It did not occur to him that I might not play chess, luckily I do (I have played for Sussex, and have a reasonably high ranking), and we had a most enjoyable three-hour battle, ending in a honourable draw. On arrival at Leningrad we disembarked, shook hands, smiled at each other and parted, never to meet again on this earthly plane. I never did know his name!

A year later I was again in Moscow, attending an international conference, and had my first meeting with Yuri Gagarin, the pioneer cosmonaut. Certainly he will never be forgotten. When he blasted off in Vostok 1, in April 1961 he was venturing into the unknown. For all he knew he might have been seared by cosmic rays, battered by meteoroids, fatally damaged by ultra-violet radiation from the Sun, or overtaken by uncontrollable space-sickness. In fact he avoided all these dangers, and made a faultless circuit of the world before landing safely in the prearranged area. We had to use an interpreter when we met, but that was no problem.

I liked Gagarin on sight. He was as pleasant as he could be, and was full of enthusiasm; I could well understand why he had been selected as the USSR's first cosmonaut. I asked him if he hoped to go to the Moon, and he gave an infectious smile. 'If I am chosen, I will go.' Sadly, he was killed in an ordinary air-craft crash in 1967, and by then the Russians had been forced to abandon the 'Moon Race'. Gagarin's first flight was the only time that he went into space.

It did not take the Americans long to recover from the initial disappointments, and they were soon matching the Soviets in almost every department. Al Shepard made their first sub-orbital 'hop', and then John Glenn completed a full circuit of the Earth. NASA was established, and President Kennedy announced that the USA meant to put a man on the Moon before 1970. During the 1960s there were any number of developments which would have sounded like science fiction even a decade earlier, but I do not propose to dwell upon them here, because my rôle was in Moon mapping and in carrying out BBC television commentaries during the unmanned missions. In particular there were the Rangers, which crash-landed on the lunar surface and sent back detailed images before committing hara-kiri; the Surveyors, which made controlled landings, and the Orbiters, which provided detailed maps of the whole of the Moon, thereby rendering all previous maps effectively obsolete. So I will merely make brief mention of some of the episodes in which I was involved, not necessarily in chronological order but all relating to the years between 1961 and 1965, when I moved house from East Grinstead to Armagh.

September 1961 marked the broadcast of the fiftieth *Sky at Night* (how long ago that seems). To mark the occasion, we decided to take a risk. Nobody had ever shown direct telescopic views of the planets on television, so why not try it, and hope that clouds would not thwart us?

Both the giant planets, Jupiter and Saturn, were visible in the evening sky, because they had both been at opposition in the previous July. (When a planet is at opposition, it is on the far side of the Earth with respect to the Sun, so that is then directly opposite to the Sun in the sky and is well placed for observation.) We needed a powerful telescope. The main instrument in my own observatory at East Grinstead was a reflector with a 12½-inch mirror; it was optically excellent, and still is, but it was not clock-driven, so that for television it was useless. As the Earth rotates, the sky seems to move round, carrying everything with it. Unless the telescope is driven so as to follow this movement, the target object will race out of view. We wanted something more elaborate, and we called in George Hole, who lived in Brighton and had a magnificent 24-inch reflecting telescope in the open air. It looked like a huge gun, but was very suitable for our purpose. Moreover, that September there was a spell of clear weather, and we felt encouraged.

We first had to work out how the television camera could be adapted to fit the telescope – something which is easy enough now, and has been done countless times, but was far from easy in 1961, when television equipment was relatively primitive. I did not pretend to be a technician in any sense of the term, but the BBC engineers were quite confident, and during the week before the programme we carried out a series of tests which proved to be satisfactory. Both Jupiter and Saturn were impressive in our monitors, despite the lack of colour (we were still in the black-and-white era, of course). Saturn, with its glorious ring system, was particularly lovely, and Jupiter's cloud belts were striking. We even saw the four large Jovian satellites, which we now know to be fascinating worlds but about which we knew little at the time.

The giant planets are quite unlike the Earth. They are big – Jupiter is nearly 90,000 miles in diameter, Saturn over 70,000 – and they are not solid and rocky; their surfaces are made up of gas, chiefly hydrogen, and the details are always changing. Jupiter shows its celebrated Great Red Spot, a vast oval with a surface area greater than that of the Earth, and which had been under observation ever since the seventeenth century. In 1961 we had no real idea what it was, and it was often believed to be a solid or semi-solid body floating in the Jovian gas. Sometimes it disappears for a while, and may be absent for a few months or even a few years, but it always comes back. Moreover, it drifts around in longitude, showing that it cannot be attached to a solid core below. In a programme before our fiftieth, I had even given a demonstration of a possible cause of its behaviour, basing myself on a theory due to Bertrand Peek, one of the most famous amateur astronomers of the time. I must digress for a moment or two to tell you what happened, because it was directly relevant to our experiment a month or two later.

According to Peek, the Spot's visibility or invisibility depended on the changing density of the gas in which it was floating. Peek had likened this to the behaviour of an egg dropped into a vase of water. The egg will sink to the bottom – but if you make the water denser by adding salt, the egg will rise to the top. Therefore, said Peek, the Spot will be visible only when the surrounding gas is comparatively dense.

I produced a vase, filled it with water, and inserted the egg, which sank. 'Now,' I said, 'I will add salt, and as the water becomes denser the egg will bob up.' I added salt – and salt – and salt; that infernal egg never did rise, though on tests both before and after the programme it worked perfectly. Don't ask me why it remained obdurate on that one occasion; I never found out!

Back to the Fiftieth ... We had a quarter of an hour on the air. With five minutes to go, the sky was brilliantly clear, and we had Saturn in view. One minute to go, and the clouds came over. We swung across to Jupiter, and the clouds followed us as if on cue. For the next fifteen minutes we swung the telescope in all directions. 'Over there, George – quickly!' 'Totally obscured.' 'Try, there, then – Saturn's visible!' 'Total obscuration.' I padded desperately, and as our transmission time drew to a close we made a final effort; the telescope hurtled down toward me, and I leaped aside just in time. We saw absolutely nothing. Five minutes later, and there wasn't a cloud to be seen. It was a classic case of what astronomers call Spode's Law: If things *can* go wrong, they *do*. (The identity of Spode, I may add, remains shrouded in mystery.)

Well, it was a good try, at least we had shown that it could be done, clouds permitting, and since then I have made several *Sky at Night* programmes from my observatory; with the 15-inch reflector, Saturn is a superb sight. But I have to admit that the Fiftieth turned out to be a comedy show, even though it wasn't meant to be.

Incidentally, we now know that Peek's theory is wrong. The Red Spot is not a solid body; it is a whirling storm – a phenomenon of Jovian 'weather' – and the colour may be due to the phosphorus compounds welling up from below. This was something which we learned much later from the spacecraft at the time of the Fiftieth, the idea of sending rockets to Jupiter seemed wildly futuristic. George died a long time ago, and his telescope is now in the Vatican. I hope the Pope enjoys using it when he can take a break from his daily task of creating new saints.

Next: eclipse chasing, to which I admit to having become addicted.

Total eclipses of the Sun are possibly the most spectacular phenomena in all Nature. What happens is that the Moon passes right in front of the Sun, blotting out the brilliant solar disk for a brief period – never as long as eight minutes, and usually much less. The Sun's diameter is 400 times that of the Moon, but by sheer chance (nothing more) the Sun is also 400 times further away, so that the two disks appear virtually equal in size. When the Sun is fully covered the wonderful corona, a pearly 'mist' marking the Sun's atmosphere, flashes into a view against a sky which has become dark enough for planets and bright stars to be seen. The whole of the solar disk has to be hidden; 99.99 per cent will not do.

The trouble is that because the Moon's shadow is only just long enough to touch the Earth, one has to be in precisely the right place at precisely the right time; the track of totality is never as much as 170 miles wide. From Britain, the first total eclipse in my lifetime was that of June 1927, when the track crossed the northern part of England. I took no part in the observations myself, partly because I lived in Bognor and partly because I was only four years old. However, I did go to Sweden to see the eclipse of June 1954, and I was suitably impressed. We were lucky; the clouds kept away.

A total eclipse was due on 15 February 1961; the track crossed

France, Italy, Yugoslavia (as it was then) and Russia. I had a luminous idea. Why not try to show totality three times, as the eclipse progressed? The Moon's shadow whips along at a tremendous rate, and my idea was simple. Station three observers along the track, at different points. Totality would occur first in France, then in Italy and then in Yugoslavia, so that if we could enlist the aid of European networks we could have several bites at the same cherry. Paul agreed, and we set up a conference in Paris to thrash out the details.

I remember that conference well. Various television representatives were there, and as soon as we met up we found that they were genuinely interested. At one stage I found myself taking the chair at a lively discussion, and I had to concentrate grimly; my French is fluent enough, but it would be only too easy to make a stupid mistake. One of the delegates was Spanish, and was very keen to join in. I had to break the news that the track of totality did not cross Spain, to which he responded 'But cannot this be altered?' Unfortunately, shifting the Earth by a few million miles would be difficult, even for the benefit of television.

We also had a rather sceptical, and formal, Italian delegation. Before lunch the discussions became somewhat sterile, and this was a pity, because Italy was a vital part of our intended chain of stations. Luckily, the Conference president, a Dane named Eddie (what his surname was I have no idea; I doubt whether I ever knew it) had the answer. He took me aside, and gave me some sage advice. 'Take these gentlemen out, and give them plenty to drink.' Actually these were not his exact words – they were couched in much more basic Anglo-Saxon – but the advice was good, and I followed it. When we returned in mid-afternoon, things were much better. Indeed, the Italian producer went so far as to suggest that the best way to time the eclipse was to let some lions lose in St. Peter's Square and see how many people they ate during totality. By the time we returned to London, everything was more or less fixed.

The resulting programme was a distinct success, particularly since it was the first time that anything of the sort had been attempted. On the other hand there were some curious hitches, and not one of our three stations worked out entirely according to plan.

We enlisted the aid of two highly experienced helpers, Dr. Hugh Butler of the Royal Observatory, Edinburgh, and my old friend Colin Ronan, the scientific historian. Hugh was dispatched to St. Michel in southern France, where there was a major observatory – fortunately, right on the track of totality. Colin went to Florence, and my destination was the top of Mount Jastrebač in Yugoslavia. Hugh would see totality first, then Colin, a few minutes later, and finally me. We did our best to set up an extra station in Russia, but this fell through, because there were no facilities available, and we were met with a polite and regretful 'Nyet'.

From England the eclipse was no more than 90 per cent obscured, but we could hardly leave ourselves out, and another old friend, Henry Brinton, volunteered to go up in an aircraft above the clouds, so that he could broadcast a summary. It was all most exciting, and as 15th February drew near we had high hopes. By this time I was already in Yugoslavia, so that it was only later that I heard full details about what happened. Paul Johnstone, in the Lime Grove studio, was in sole charge.

Henry opened the programme from 20,000 feet. This went well, and although we could show no pictures the commentary was excellent. Then the centre of attention switched to St. Michel, where Hugh Butler was ready and waiting.

This was Hitch No. 1. The cameras went 'live'; there was the sun, and at the appointed time totality began, with the sky darkening and the corona flashing into view. Of course, this was in monochrome (colour TV lay in the future), but it was dramatic none the less. Unfortunately no commentary came through. Hugh was seen on the screen, talking in animated fashion, but not a sound could be heard. Subsequently we found out why. Hugh's commentary went from his mouth to the cameras, from the cameras to the Observatory, from the Observatory to the main transmitter in Milan, from Milan to London, from London to Lime Grove, and thence to the studio – where everything would have been fine if only the technician had remembered to plug it in. With admirable presence of mind, Paul switched to a stand-by commentary, and prepared to go over to Florence.

This time we had both sound and vision, with Colin giving a graphic description of the scene as totality approached. 'The light is fading,' he said. 'It's impressive, and there is a deathly hush everywhere.' At that moment the last segment of the Sun vanished, and the assembled crowd let out a 'Waaaaaah!' which could probably have been heard in Lime Grove even without the help of a transmitter.

Over in Yugoslavia I was having problems of my own, and because of the failure of my radio link with Milan I was completely out of touch. I had no idea whether or not we were on the air, or whether either of the first two broadcasts had been successful. I felt distinctly isolated.

Mount Jastrebač is a fairly high peak, not far from the town of Niš (pronounced Neesh). I was the only Englishman in the part, though there were various astronomers from other countries, some there as broadcasters and others as pure scientists. The reasoning was that the cloud-level was likely to be low, so that if we went to the mountain-top – where there is a small radio station – we would probably have a clear sky. In fact the cloud-level was high, and we were immersed in murk. There was also the complication that the Yugoslav director was a man with Ideas, about which I was blissfully ignorant. Our equipment had been taken up the mountain trail in carts pulled by oxen (as least in theory; so far as I can remember we spent most of the journey pushing the oxen) and the director decided to introduce a charming Nature note. It is said that when the sky darkens at totality, animals are fooled into believing that night has fallen, and go to sleep.

Therefore, why not turn the cameras on to the oxen at the critical moment, and show them dozing peacefully off?

I was unaware of this, because of linguistic difficulties. I had to talk French to a Belgian astronomer, who relayed it in German to the Yugoslav director, who in turn passed it on to the cameramen in Serbo-Croat. At the preliminary discussions we had literally stood in a circle, playing what was once known as the children's game of 'Whispers', and it was cumbersome by any standards. I only hoped that it would work.

Five minutes before totality we were still in cloud, and it looked as though we would see nothing at all, but at least it would be obvious when totality started, because the light-level would drop abruptly. It had been arranged that I should begin my commentary three minutes early – by that time, totality in Italy was over – and all I could do was to start talking and pray that I was being heard. As I later found, Paul was to start talking and pray that I was being heard. As I later found, Paul was staring at the screen, more or less gnawing his nails, and hoping that I would appear on schedule. To his immense relief, I did. Unfortunately the Sun didn't; the cloud-cover was complete, but there were signs of an approaching break, and I calculated that if all went well it might arrive in time.

Then, suddenly the light faded. 'Totality is almost on us, 'I said, 'and everything is getting very dark...' It was then that the director switched the cameras on to the group of oxen. Just to make sure that everyone could see them properly, he floodlit the brutes. Naturally they simply chewed the cud (if that is what oxen do) and looked silly. I made a gesture which could not be misinterpreted even in Serbo-Croat, and the cameras turned back to the Sun.

We were fortunate. The length of totality was less than three minutes, but with about fifty seconds to go the clouds broke; there was the corona, looking glorious, and the effect was breathtaking. As the Moon edged away from the solar disk we saw the brilliant flash of what is known as the Diamond Ring – the first sliver of sunlight reappearing – and then the light flooded back over the mountain. Almost at once the clouds closed in once more, but I felt well satisfied.

I ended my broadcast five minutes later, still without the slightest idea of whether I had been heard. It took us several hours to pack up, and several more to make our way down the mountain down to the little village at the bottom. En route, I made an unwise decision. Everyone was going to ski down from the summit to the half-way hut, and I elected to do likewise, so I borrowed some skis and set off with the rest. I ought to have known better; as we came in my left knee, badly smashed up when I was twenty, gave way. Luckily I wasn't really hurt, but it did force me to realize that so far as I as concerned, skiing was 'out'.

On arrival we went into the only pub, where we celebrated by drinking a great deal of slivovitz, the local plum brandy, which is not nearly so harmless as it looks. Finally we saw what we thought were two jeeps coming to take us back to Nis. In fact there was only one, but it served, and I was able to call Paul Johnstone, who told me – to my relief – that the programme had been carried through.

At the time it was traumatic, but at least it was a new experiment for television, and was also the first of our really ambitious *Sky at Night* programmes. Since then we have covered several more eclipses, but I will always have fond memories of that one – particularly the sight of those darned oxen blinking stupidly in the glare of the floodlights.

Two teenagers came into my orbit at around this time: Peter Cattermole and Iain Nicolson, who became, and have remained, very special friends – and it was they whom I called upon for help at a very sad moment in my life, of which more anon. Peter was studying at the East Grinstead Grammar School, and intended to make his career in geology; now, as Dr. Peter Cattermole, Principal Scientific Investigator for NASA on the subject of Martian volcanoes, he has made his mark in no uncertain manner.

Unexpectedly, I had a call from *The Times* newspaper. There had been reports of the fall of a meteorite at Morsgail, in the Scottish island of Lewis; would I go and investigate? Looking up the records, I found that meteorite falls had been reported there before, and that showed me that we were dealing with nothing cosmical, because meteorites do not fall in the same place. It had to be geological. I called in Peter, and we set off in my Ford car, the Ark, which when pressed can reach 40mph on a level road, though it is not so rapid uphill (once, going up Bury Hill near Arundel, I remember that a spaniel scuttled past me).

Terrestrial navigation was not our forte, and we were baffled when we decided to take photographs of the Forth Bridge; we failed, because at that time we were actually crossing the Clyde. We took an airlift to Lewis, and drove around in the smallest car I have ever come across; to use the clutch I had to take my shoes off, and sheep were constant menaces. Finally we located the object, which predictably, turned out to be a rock of the type known as a geological erratic. It had not descended from the heavens, and what had apparently happened was that a methane explosion had blown the rock from a low-lying pond in a boggy landscape up into a depression at a higher level. As I remember saying at the time, we had come a long way merely to report on a bang in a bog.

I went twice to East Germany, to take parties round the Zeiss works in Jena, where the first planetaria were made. On the second occasion I had Iain with me. Frankly, the German Democratic Republic was a sinister place, and I liked it even less than West Germany. We were given an interpreter who had one obvious fault: she spoke only German. Her name was something like Mrs. Floorpolish, and though she did her best she was not a great deal of help. Our party totalled about twenty, and her sole contribution was to say 'I vill ask'.

The Zeiss factory was most interesting, and so were the observatories we managed to visit en route. We came back through Checkpoint Charlie, in the Berlin Wall, where quite a number of would-be escapers had been shot. Mrs. Floorpolish had said goodbye (as an East German she was not allowed anywhere near the Wall), and I was in charge. Armed Germans brandished guns as we drove into checkpoint, and spoke volubly; I used Semaphore to explain that I spoke no German, and eventually they produced another German who could speak French. I admit that I could not resist the temptation to take a photograph. If the Germans had seen me doing this they would have been most annoyed, but I made sure that they didn't, and my knowledge of that sort of procedure was rather greater than theirs.

By that time I looked rather sinister myself, because I was horribly unshaven. Iain had dropped my one electric razor when we were in Leipzig, and there was no time to buy another; it did not affect Iain, who at that time was barely old enough to shave at all – though he subsequently followed Peter's example and hid himself behind a luxuriant beard.

We met no astronauts during that trip, but in 1966 Valentina Tereshkova paid a visit to London, and I was asked to take the chair when she addressed a large audience in the Royal Festival Hall. Her English is fairly good, and she is utterly charming; as you may remember, she was the first woman to go into space. At that meeting, a very tough journalist from (I think) the *Daily Mirror* stood up and asked: 'What qualities would you look for in a man going on a flight to the Moon?' Valentina gave him a delightful smile. 'Do you mean if I was going too?' Collapse of journalist.

Yes, life was busy, and it was difficult to find time for everything I wanted to do. Even weekend cricket had sometimes to be put on hold, and this pained me, notably when our wicket showed signs of taking spin. But there were additional problems to be faced, and these, to my regret at the time, led to a move from the East Grinstead house where I had lived for thirty-six years. My whole lifestyle was about to change.

7 Irish Interlude

I had no wish to leave England, and in particular no wish to leave East Grinstead, which was 'home'. Unfortunately there were real problems to be faced. I will gloss over them, because they are of absolutely no interest to anyone except me, but for the sake of continuity I cannot omit them entirely.

As I have said, I was exceptionally close to my mother. Before the 1914 war she trained as a singer in Italy, under Sabatini and Clerici, and even before she had finished her training she was offered a soprano lead in Italian Grand Opera. But when war broke out she naturally came home; she married my father (an Army officer), and never actually did any professional singing. She was also a talented original artist, and her paintings of what she termed 'bogeys' were magnificent. At the age of 87 she published her one and only book, *Mrs. Moore in Space*.

I wish I had inherited those gifts. Apparently I could sing well when I was a boy, but when my voice broke it didn't just break – it shattered. And if I had the slightest spark of artistic ability I would try to develop it; alas, I am so hopeless that there really isn't any point.

During my boyhood my mother had to shoulder virtually all the main responsibility, not helped by the fact that my father was never fit, and I was a crock between the ages of six and fifteen. The East Grinstead house was not ours – it had been built for her, and was joined on to the next door house by a closed passage. This was for the benefit of my maternal grandmother, who was an invalid and needed someone close by – she was a delightful person with a vast sense of humour. She died in 1939, and a complicated will meant that although Mother was a life tenant of the house, I wasn't. When she died, I would have no home. This did not worry me in the least, but it did worry her, so that clearly a solution had to be found. I did not have enough money to buy the house, and so when the offer of a really interesting job came along it had to be taken seriously. I had never had a full-time job, so that I had no settled income and no guaranteed pension. Being a freelance is great fun, and the only life for me, but it does have its disadvantages. Had there been any prospect of my getting married, I doubt whether I would have been able to stay unemployed.

The offer came from Northern Ireland. Would I like to be Director of a new planetarium to be set up at Armagh?

Armagh, well inland from Belfast, is a fair-sized town, and was the centre of Northern Ireland's astronomy because it contained the only professional observatory in the province. That observatory was old, and had a distinguished record, though the equipment was limited, and the main telescope was a 10-inch refractor. The current Director was Eric Lindsay, whom I knew well. He was a world-famous astronomer, and in the best sense of the term he was also a publicist. He was anxious to give Armagh a planetarium; as a fund-raiser he was amazingly skilful, and he persuaded the Belfast government (Stormont) to provide an adequate grant. So he went ahead, and contacted me as a possible Director, because he remembered that I knew something about planetaria.

In fact my connections were decidedly nebulous. When the London Planetarium was set up at Madame Tussaud's, in 1961, I was invited to become Director. I said 'no', partly because I had no wish to live in London and partly because at that time I had no need of a full-time job; I was much too busy. When I turned them down, the Planetarium authorities appointed a Dr. H. C.

King. I have to admit that he and I were not on the same wavelength at all, and there was another curious episode which was relevant. I had been asked to write the first Planetarium brochure, and I did so; R. E. Edds, of the Planetarium board, received it very enthusiastically, and thousands of copies were to be printed. Then I was asked to make a few amendments, in view of new information. I sent the manuscript back, and Edds wrote to me: 'You have ruined it; we can't publish this.' In fact I had not altered one word. When the brochure was published, my name was conspicuously absent. I merely laughed, and opted out.

However, I had given displays in other planetaria, notably America and South Africa, so that I was not a total beginner, and Eric's suggestion was not unattractive, despite my reluctance to be tied to a job. Of course, I would have to be in London regularly, for *The Sky at Night*, but the air links seemed reasonable. Mother and I talked it over again and again; left to myself I would probably have stayed in East Grinstead, but she was still very worried – she was in her seventies, of course – and in the end we decided to go. Breaking up what had always been my home was something that I did not enjoy in the least, but it had to be done, and in the summer of 1965 we arrived in Armagh, where I had bought a house – the old judge's house, in the Mall.

Despite my Irish name, I have very little Irish blood in me, at least so far as I know. Certainly I had no Irish connections apart from Eric Lindsay, and I knew little about the situation in Ulster. Things were quiet when we arrived, and in fact this was the most peaceful period for a long time; though we knew that the IRA existed, it did not seem to be much of a threat. Had I realized the true depth of bitterness between the Protestants and the Catholics I would have stayed firmly in Sussex, but I hadn't. As soon as I arrived I could feel the tension, and I resolved to have nothing to do with it.

I did not understand the problem, and I still don't. After all, both sets worship the same God; the God of the Vatican is the same as the God of the Protestant leader, Ian Paisley, whose malign influence was (and is) evident. If the Church leaders were sincere, they would work together instead of fighting like Kilkenny cats. In Armagh there were two Archbishops, one Protestant and the other Catholic. During my stay I met them both, and I remember thinking that the best course would be to take those two silly old coots and bang their heads together.

The judge's house – 4 St. Mark's Place, The Mall – was old and large; it had three floors, and also cellars. It was known that all the Mall houses had rats in the cellars. On arrival I set a cage trap, and on the following morning it contained a grey, rather nice-looking rat. I had engaged an odd-job man. 'How are you going to kill it?'

'Leave it to me.' I put the cage in the boot of the car, drove out into the country, and deported the rat. We parted with mutual expressions of esteem; it was a handsome animal, and I had no wish to harm it, though I certainly did not want it as a house guest. The word must have got around in the rodent colony, because I never saw another rat. When asked how I kept them away, I merely replied that I had my own methods. They may have been unconventional, but they did work!

Several episodes come to mind about the first few months in Armagh. First, the cricket club encounter. I was a keen cricketer (of this, more anon) and one of my first acts was to go down to the local club, with the set intention of joining. I met the Secretary, who asked me several questions.

'Are you a bat, a bowler or an all-rounder?'

'Purely a bowler; leg-spin, medium pace. No. 11 bat.'

'Good; we need a spinner. Are you Protestant or Catholic?'

I looked at him in amazement. 'What on earth does that matter?'

'Of course it matters here. It makes a lot of difference.'

I rose to my feet. 'I'm a Druid. Good afternoon.' – and I never went near the Pavilion again. I rather think that it was this incident which made me realize that my stay in Northern Ireland would not be prolonged.

My study in No.4 was on the ground floor, and looked straight out onto the Mall. I bought a statue of Buddha (or some such deity), put it on my windowsill, and went out early each morning to pay my respects to it, knowing that I was being watched. Everyone was agog to know which Church I attended, and on finding that I attended neither there was considerable bewilderment. It was weeks before the locals realized that they were having their legs pulled.

Much more significant was my connection with the Scouts. At East Grinstead I had been active in the Scout movement – not in uniform, but as a 'helper', so that when I went to Armagh the Scout group there asked if I would become County Secretary. I said 'yes', and all went well until the time of the annual Three Counties Conference, comprising Armagh, Tyrone and Fermanagh. As County Secretary for Armagh, the host town, it was my job to organise speakers, venues and much else. Behind my back, the Protestant Scouters met – not in my study, as usual – and told me that no Catholic groups could attend. You can imagine my reaction to that. I stormed into Belfast, got nowhere, and then flew to London to see the Chief Scout, who had given me a Thanks Badge some time earlier. He was sympathetic, but said, sadly, 'I know how you feel, but we can't fight the Church.' I handed him back my Thanks Badge, and that was the end of my official links with Scouting.

However, it had brought me into touch with a teenage Scout, Terry Moseley, who became (and remains a close friend). Today he is a senior Civil Servant in Belfast, but his hobby has always been astronomy, and he is a far better observer than I could ever hope to be. Mind you, he is blessed with telescopic eyesight!

When I arrived in Armagh, and settled in, the Planetarium building had not even been started, and the site was merely a grassy field in the Observatory grounds. My first duty was to choose a projector, and there were three main possibilities: German (Zeiss), American (Spitz) or Japanese (Goto or Minolta).

Me - aged about three. Do you recognize me?

W. S. Franks, who ran Brockhurst Observatory at East Grinstead. Photo taken around 1933.

Brockhurst Observatory with Me, acting Director, aged 14. Taken with my camera, 1937.

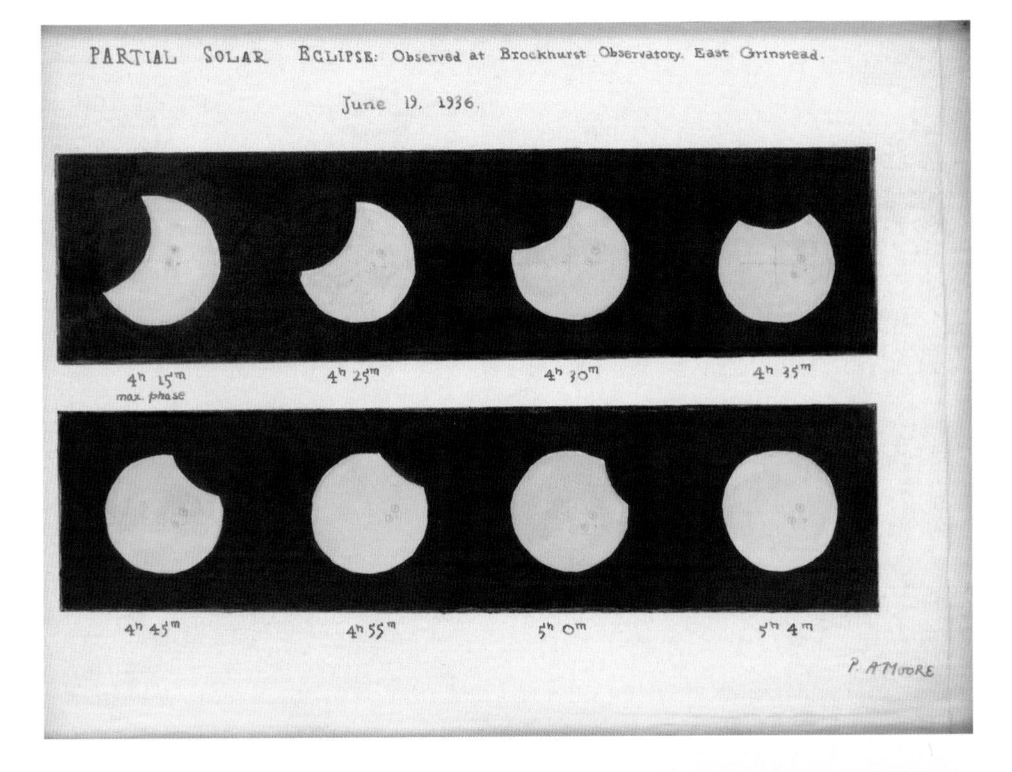

Partial Solar Eclipse.
Observed at Brockhurst
Observatory, East Grinstead.
June 19, 1936.

The Kid, aged 17!

Mother, holding our beloved Rufus; about 1946.

Remnant of Herschel's largest telescope, the 40ft reflector. I took this at Observatory House, Slough, just before the House was pulled down.

With Frank Hyde, the radio astronomer, at a BAA Meeting around 1955.

The Ark (left) and my MG Magnette (right) at Selsey, 1969. The Ark still runs!

With Neil Armstrong, in 1970. Back from the Moon!

On TV with Eric Morecambe. I am not too sure of the date!

Apollo broadcast, 1970, with Geoffrey Eglinton and Garry Hunt.

Me in a space-suit, with a very young Lawrence (left) and Matthew (right).

The Japanese were efficient, as they always are, and contacted me. They invited me to go to the United States, and examine the installations there. I did so, and it was clear that the Japanese's projector was the best for us. It was cheaper than the Zeiss, and just as good – much better than the American Spitz, where the projector looked remarkably like a large lavatory cistern. When I had made up my mind, the Japanese asked me to go to Tokyo and design some new star-plates for them.

I did so – and here I have to admit that I am totally illogical. If I go to Germany, the sooner I get out the better. I do not have the same feelings about the Japanese, simply because I never came across them to the same extent; in fact the only Japanese with whom I had much to do was the Japanese-American who taught me unarmed combat – (and boy did he throw me about!).

Japan was interesting, and I have many memories about that trip. I had dinner with some leading Japanese's astronomers; we began with raw fish, octopus, and hundred-year-old eggs dipped in Soya sauce, washed down with a great deal of Sake (Japanese's rice wine). I enjoyed it immensely, and Professor Miyamoto commented: 'Ah! you are good Japanese!' We also went to a small village in Hakone, the volcanic areas north of Tokyo, and I had to accept local customs; on entering your hotel you have to remove your shoes, and I was acutely aware of a large hole in my sock. Moreover, my shoes are not small. When ordering a new pair, I have to contact the nearest shipyard. For dinner, you change into Japanese robes; on me, the kimono looked like a ballet-skirt, and also disclosed the fact that I am violently knock-kneed, so that I could not possibly stop a pig in a passage. One of the diners asked me if he could take a photograph. I had not the slightest objection, but I wish I had remembered to ask him for a copy!

What can you make of the Japanese? Courteous, effective, orderly, welcoming; and yet during the war they were every bit as barbarous as the Germans. I give up.

When I got back to Armagh, having done all I set out to do in

Tokyo, the Planetarium building was under way. I also had to cope with office equipment, and I recall one episode. A sales representative called on me to demonstrate a photocopier. We plugged it in, switched on – and it burst into cheerful flame. There was a great deal of paper around, but the fire extinguisher hadn't arrived. Luckily we had been having a pre-lunch drink, and on the table was a soda-siphon, so I grabbed it and turned it on to the holocaust. It did the trick, but also drenched the salesman. When the fire was out, he looked rather rueful. 'Shall I come back tomorrow with another model?' I said that I didn't want to bother him!

There were problems with the dome, but by the end of the next summer they had been sorted out; the projector arrived, and was installed, so that we could begin regular shows. The Planetarium staff consisted of myself and a handyman, so that I had to do all the lecturing; I enlisted some helpers, notably Terry Moseley, but it was very much a full-time occupation, and apart from *The Sky at Night* everything else had to go 'on hold'.

I did make good use of the 10-inch telescope, and observed regularly, usually with Terry, whenever the sky was cloudless (which in Armagh is not very often). One episode involved a past Observatory Director, Dr. Davenport, who had committed suicide in the Director's study about a hundred years earlier and was said to indulge in a little casual haunting. One evening Terry was alone on the dome when he heard footsteps coming up the path, and assumed that I was on my way; at the time we were busy making systematic observations of Jupiter. Nobody arrived, and it was only half an hour later that I came up, because I had been delayed by a lengthy phone call. Had there been a visit from Dr. Davenport? We will never know!

The Planetarium was a great success, and I had many shows per week, usually to full houses (the seating capacity was 150). Yet from my point of view the situation was not ideal. I was always conscious of the uneasy atmosphere, and of course I had to go to London regularly for the *Sky at Night* and other television programmes.

I used to fly from Belfast to London, initially on BOAC and then on BUA (British United Airlines), who were far better. BUA, unlike BOAC, used to provide reasonable in-flight meals. (The usual BOAC fare was a sandwich made at the time of Edward VII.) On one occasion I was flying over for a programme, on BUA, and bit on what I thought was a grape. In fact it was an olive. There was a nasty snapping sound, and there I was, with two halves of a broken denture. (For reasons which need not concern us here, I have had complete dentures, upper and lower, ever since the age of nineteen.) It was a Sunday; I was due on the air, live, in a matter of hours, and there was not the slightest chance of finding a dentist. There was only one answer: Sellotape. I don't think that that was one of my better broadcasts, because of the danger that my dentures would fall out, but I survived.

On arriving back in Armagh, where (mercifully) I had a spare set, I had a phone call from the BBC. 'We can't find our best stop-watch. Did you see it, by any chance?' I disclaimed all knowledge of it – until next day, when I opened my night bag and found a stopwatch inside it. I rang the BBC. 'If you look in your stopwatch case, I'm afraid you will find a set of broken dentures.' That took a great deal of living down!

Astronomically, one event of 1967 was the discovery of a nova, or new star, by the English amateur George Alcock. A nova is not really a new star at all; what happens is that a formerly faint star suffers an outburst, and flares up to many times its normal brilliance, remaining bright for a few days, weeks or months before fading back to obscurity. George rang me at about two o'clock in the morning. 'There's a fourth-magnitude nova in Delphinus. Dash out and confirm it!' He gave me the position; I rushed out – and there was the star. It was visible with the naked eye, easy in binoculars, and I could identify it at once, though I would never have done so had not George told me exactly where it was. I was in fact the first to confirm Nova HR Delphini 1967, though I can claim absolutely no credit.

Less successful was the case of the tiny inner satellite of Saturn. Saturn, the Ringed Planet, has a whole family of moons; only one (Titan) is really large, and some of the others are very small and close in. They are best seen when the rings are edgewise onto us, as happens every 15 to 17 years. It did happen when I was at Armagh, and, with Terry, I made a long series of observations of the elusive inner satellites. I did not realize that one of these was new, and I found out only when it was announced later. Nowadays we call it Janus. So a satellite slipped through my grasp!

There is a great difference between the two Irelands; the ridiculous religious bickering in the North is virtually absent in the South, and during my spell at Armagh it was always rather a relief to drive across the Border. My link there was the Birr, some way from Athlone, once the home of the largest telescope in the world – the 72-inch reflector, made by the third Earl of Rosse, and completed in 1845.

It was a most extraordinary telescope, unlike anything made either before or since. The mirror was made of metal; the huge tube was hinged at the bottom, and the telescope swung between two massive stone walls, so that it could swing to and fro by only a limited amount. Yet with it, the Earl discovered that some of the fuzzy patches called nebulae are actually spiral in form, like Catherine wheels. We now know that these spirals are external galaxies, thousands of millions of light years away, and containing up to a hundred thousand million stars each. For more than half a century the 'Leviathan' was in a class of its own.

The Third Earl was followed by his son, the Fourth Earl, who was also an astronomer. When he died, in 1909, the telescope was used no more; the mirror was taken out, and the tube was left lying, forlornly between its walls. This was the situation when I first saw it, in 1967, when I was a guest at Birr Castle and was staying for a social weekend; the Sixth Earl (Michael) was not an astronomer, but he was keenly interested, and had taken care to preserve all the records.

Over dinner one evening, I had a great idea. 'Let us get that telescope back into use!'

It sounded mad; there was so much to be done, but in the end we managed it. It took thirty years, but today the telescope is fully operational, and looking just as it must have done in 1845. Of course, it cannot claim to be the equal of a modern telescope of the same size, but it is unique, and its place in history is assured. I am proud to have played a part in its restoration.

Meantime, the Armagh Planetarium flourished. I began to feel that I had made all the contributions of which I was capable, and Sussex called. In Ireland we felt isolated, and neither Mother or I liked the tension, though the major IRA flare-up did not happen until we had departed.

Eric Lindsay did not want me to go, but he did understand how I felt, I simply did not fit happily into the Northern Irish scene, and I had seen the Planetarium well and truly established. We cast around for a new Director, and approached Tom Rackham, an astronomer who was working with the great radio telescope at Jodrell Bank in Cheshire. Tom agreed to come; Terry Moseley held the fort during the interregnum, and Mother and I made ready to go home.

I had always wanted to live in Selsey, and so I flew over house-hunting and stayed with Henry Brinton, who was a tower of strength. I found Farthings, old and thatched, and fell in love with it. It had not been lived in for four years, and was going at a give-away price; the weeds in the garden were (literally) six feet high, and the thatch needed repairing. I asked two questions. 'Has it got wet rot, dry rot or woodworm?' No. 'Has it got drains?' Yes. And I bought it. It was a colossal financial gamble, because I had no reserves, but we decided to take the risk.

There was one final crisis. I sold No. 4 St Mark's Place to a local vet, who seemed at first to be perfectly decent. After the contract was signed he claimed that the Observatory, which I had set up in the garden, was now his property. I was fairly safe, because the Observatory was not fixed down, but I was not happy. He made his claim at nine o'clock in the morning. By teatime the Observatory had been dismantled, and it and the telescopes were on their way to England. There was nothing that the vet could do about it, and I do not know how he reacted; our paths would never cross again.

So goodbye, Northern Ireland. My close friend Ian Baker, who had helped me to move over, flew across to help me to move back, he drove my MG car, while Mother and I followed sedately in The Ark. We arrived in Selsey on 29 June 1968, to find that all our furniture was still on the high seas, held up for some reason or other. But at least we were home.

Looking back now, I can see what a gamble it was. I had bought the house without a mortgage; I had paid for the move, and I did not owe a penny in the world. But my bank balance stood at exactly half a crown.

8 Moon Missions and Moon Men

I was back in England before the IRA sparked off the new phase of violence in Northern Ireland. I have been to Armagh a couple of times since, and of course I have been a frequent visitor to Birr, but otherwise my connections have been decidedly slender.

At Selsey there was a long settling-in period, and there was a great deal to be done; clearing the house, dealing with the weeds in the garden, checking on the essential services... Farthings was partly thatched, and I was under some pressure to have the whole roof tiled. But I like thatch; it is cool in the summer and warm in the winter, and it looks great, so we found a thatcher and had the entire roof re-done (it took the best part of a year). Squirrels were a problem, because if they can get in they are very obvious; they sound as if they are playing football in the loft – and our loft is small and fairly useless. I would never hurt a squirrel, but I coped by investing in an electric scarer which emits a high pitched note that they do not like, though it is quite inaudible to humans. Thereafter they kept away from the loft, and merely looked at me reproachfully as they munched nuts in the garden.

Re-decoration was needed, but at the outset there were financial problems, as my half-crown reserve did not go very far. What saved me was Apollo 11, the spacecraft which carried Neil Armstrong and Buzz Aldrin to the Moon; on 21 July 1969 they stepped out onto the waterless Sea of Tranquillity. A few weeks earlier James Mitchell, director of the publishing firm of Mitchell Beazley, had asked me to write a *Moon Flight Atlas*. I said 'yes', but I doubted if it would make much money, because it would sell no more than a couple of thousand copies at most. I wrote the book at breakneck speed, it was published within a month of the lunar landing – and it sold 800,000 copies putting me on my feet.

When the first cash from the book came through, I asked Mother what she would really like done. Her answer was: good central heating. The system we found on arrival seemed to be Roman; you lit it with a match, while most of the house was heated by a curious solid-fuel burner which stood in the middle of the main kitchen, and periodically exploded, giving off poisonous gases. We had both systems ripped out and replaced with oil heating, later augmented by solar panels in the roof. The results are very satisfactory. But for Mr. Twitmarsh, I might have opted for gas – but more about Mr. Twitmarsh below.

The garden is large, and attractive; it contains my observatories, and we also have a croquet lawn, even if it is not quite up to international standard. There are various local rules. For instance, if you hit a ball into the shrubbery you forfeit a turn, and sending a ball under the garden châlet means that you have to go back two hoops; the player using the yellow ball has an extra turn, because the yellow ball has a chip in it and rolls erratically. It is fun; as you may know, croquet is a vicious game, and about the only one in which men and women can take part on genuinely equal terms. I understand that it has been responsible for more divorces than any other game ever played.

On arrival I was asked to become one of the Meteorological Office's local weather stations, and of course I agreed; I am station No. 530000770. Usually we have good weather, and indeed a sort of mini-climate which includes Selsey, Bognor and not much else, this was why King George V went to Bognor to recuperate, turned it into Bognor Regis, and made that unfortunate remark about it. But we do have our moments, and a few years ago Selsey was struck by a full-scale tornado. It was quite terrifying; it passed between my house and the main observatory, knocking down a wall and dislodging the roof of the smaller observatory. Had it been a few yards further east, it would have hit the house and stripped the thatch off.

I was nearby. A friend of mine had arrived by car at around 11.30pm (he had been held up in a massive traffic jam), and I had taken him to the local Indian restaurant to have a curry. Just before midnight, all hell broke loose. The door bulged in, and I realized that the drop in pressure indicated a tornado; we made for the centre of the restaurant, and waited. As soon as things calmed down we made our way down the road, half expecting to find a flattened house. In fact we had escaped serious damage, and though the tornado whipped right through Selsey nobody was hurt. Subsequently the Indian restaurant was re-named 'The Selsey Tornado!'

My life between the move to Selsey and the end of 1972 was totally dominated by the Apollo programme. Really it began in December 1968, when the astronauts of Apollo 8 flew round the Moon and became the first men to have a direct view of the far side. It ended four years later, with the return to Earth of Apollo 17. I will say little about the missions themselves, because other books do that far better than I could; I will concentrate solely upon the aspects in which I was personally involved.

First and foremost, the Moon-men. I have often been asked what singles them out from the rest of humanity. My answer is always the same; as individuals they differ in personality, outlook and lifestyles, but all have more than their fair share of courage, ability and resource. Only twelve men have so far walked on the lunar surface, and they are very special people. The same is true of the astronauts who have orbited the Moon without touching down there.

For example, consider the first two Moon landers, Neil Armstrong and Buzz Aldrin. They are not in the least alike. Neil is not fond of publicity, and in general is not too keen on being interviewed; Buzz, on the other hand, thrives on publicity, and is extremely good at it. He has lectured all over the world with great effect, and has inspired at least two generations. Neither is he afraid to speculate; he is looking ahead to tourist trips not only to the Moon, but even to Mars. Full-scale space tourism may be delayed for several decades, so that Buzz may not see it – but I hope he does; he deserves it, and would indeed make a magnificent guide.

Incidentally, let me nail another rumour which has been quite widely circulated. Most of the photographs from the Moon during the Apollo 11 landing show Buzz, not Neil, and the story goes that this was deliberate on Buzz's part, because he was piqued at being the second man on the Moon rather than the first. I have never heard such nonsense, and indeed Buzz would be the very last person to act in such a way. The reason why Buzz is shown is because Neil was taking the photographs – and when hampered by a space suit and limited in time, one cannot bother about changing cameras. I hope that we have heard the last of this peculiarly nasty rumour. Incidentally, Aldrin's name really is Buzz; he was christened Edwin, but changed it by deed poll.

Of the other Moon-men, some are religious, others not; some are thoroughly at home on radio and television, others less so – but all retain the same enthusiasm and inspiration. I say 'all' though, sadly, two – Jim Irwin and Al Shepard – are no longer with us.

When I arrived in Selsey the scene was set for Apollo, and on 21 December 1968 Colonel Frank Borman, Captain James Lovell and Major William Anders, in Apollo 8, blasted off from Cape Canaveral, and by Christmas Eve they were nearing their target. Their first task was to fire the spacecraft's rocket motor and put themselves into a closed path round the Moon. This had to be done when they were actually *behind* the Moon, so that they were completely out of touch. Their radio signals could not reach Earth, and they were cut off from all mankind.

At that time most of the news broadcasts were transmitted not from Lime Grove, but from Alexandra Palace, near Wood Green. It was here that I was stationed, on my own apart from the BBC technicians; everyone else had gone home. I was on the air 'live' as the critical moment approached, and I can more or less remember my exact words:

'The Apollo 8 astronauts have passed behind the Moon, and by now they will have fired their on-board rocket, with the aim of putting themselves into lunar orbit. But remember, there is only one chance. If all goes well, we will hear their signals in less than one minute from now. If the rockets have misfired, there is a chance that they will have hurled themselves into the wrong orbit – and of course there is always the dreaded possibility that we will not hear from them at all, in which case we may never know what happened. All we can do is to wait. I will say no more; listen out for the signals in less than twenty seconds. This is one of the great moments in human history.'

And the BBC switched over to *Jackanory*.

I did get back on the air a few minutes later, by which time, thankfully, the Apollo signals had come through on schedule. But it was an episode which must surely remain unique in the annals of broadcasting!

Apollo 9, next in the series, was an Earth orbiter designed to test out the lunar module. Apollo 10 was another circumlunar flight, and then, in July 1969, came Apollo 11. I was one of the two main commentators on BBC television (James Burke was the other), and we remained so up to the end of the Apollo missions in December 1972.

As the Apollo missions progressed, the BBC coverage became more and more elaborate, and this led me on to formulate what is still known as Moore's Law: 'The efficiency and interest of any television programme varies in inverse ratio to the number of producers and directors involved'. For Apollo 8 to 11 we had one producer, one director and two commentators (James and me).

By the end of Apollo 17 we had at least ten producers and swarms of commentators, though James and I still acted as the link men. Of course, some of our studio guests were first-class broadcasters as well as being eminent scientists, and I have particular memories of Dr. Stuart Agrell and Professor Geoffrey Eglinton, the lunar geologists, who made splendid contributions. They usually occupied one corner of Studio 7 in Television Centre, and I am sure they will not mind my recalling that we used to call their contributions the Agg and Egg Show!

Apollo 11 was launched on 16 July 1969. A week earlier the Russians had dispatched their unmanned Luna 15, with the aim of landing in the Sea of Crises and bringing back samples of Moon rock (something which had not been done before, though it has been achieved several times since). Nobody was sure whether or not Luna 15 was an attempt to upstage Apollo. I was convinced that it wasn't, but all the same the BBC asked me to fly to Moscow and see what I could find out.

This was quite a challenge, because my knowledge of the Russian language is nil, and when I arrived I found that the resident BBC staff could not be of much help. Then I had a stroke of luck. An astronomical conference was being held at Moscow University, and as a member of the Internal Astronomical Union I was able to obtain an invitation. I then managed to collect three English-speaking Soviet scientists, a camera crew, and a sound recordist on the University steps at the same time, and returned proudly to London with a film which was shown that night. I am convinced that the Russians were merely continuing with their own missions in their own way; actually Luna 15 failed, as it crash-landed on 21 July and sent back no useful data. By then, naturally, all the world's attention was concentrated upon Apollo 11.

One of our minor problems was air-conditioning. Studio 7 in Television Centre is large, but the weather that week was swelteringly hot, and the lights raised the temperature to a positively

Venusian level. We had a large globe of the Moon suspended in a prominent position, and every time we switched on the air-conditioning the globe danced around in a most bizarre manner, so we had to perspire and make the best of it.

On the night of the landing itself I was broadcasting continuously for more than ten hours, and I will not forget it; by the end I was a little weary. The main weakness of the Apollo project was the lack of any provision for rescue, so that if the capsule made a faulty landing, or if the single ascent engine of the lunar module had refused to fire, the astronauts would have been doomed. As Neil Armstrong and Buzz Aldrin were descending in the *Eagle*, the lunar module of Apollo 11, I could not help thinking: 'Suppose we've all been wrong! Suppose the ground is unsafe, after all...' When I heard Neil's voice coming through – 'The *Eagle* has landed' – I felt a tremendous surge of relief, shared, I have no doubt, by the millions of people who were viewing or listening. I do not remember quite what I said, but I hope it was appropriate. I can never check, because the BBC, with superb efficiency, has lost all the tapes.

We stayed on transmission for Neil's 'one small step' on to the Sea of Tranquillity, and I did not come off the air until after breakfast on the following morning. Mercifully Joan Marsden, the floor manager, presented me with what looked superficially like a cup of coffee, but which was, I think, almost neat brandy. I am glad to say that it did the trick.

Each Apollo mission had its own characteristics. With No. 12, in November 1969, the lunar camera was put out of action almost at once because it was accidentally pointed at the Sun, with the result that we had to carry out our commentaries more or less 'blind'. Apollo 13 was, I think, the most tense, for obvious reasons. I had gone home after a broadcast when, at 3.00 a.m., I was telephoned with the news that there had been an explosion on board the spacecraft, and I was required. I drove back to Shepherds Bush at a furious rate (for the Ark, I mean; at times, when coasting downhill, I must have touched 45 m.p.h) and stayed there, broadcasting frequently, until the crisis was over. At one point I was able to talk (privately) to the controller at NASA, and asked: 'Are you going to get them home?' The answer: 'Yes, if nothing more goes wrong, but we've come to the end of our resources.' Mercifully all was well in the end. For a brief period it seemed as though the nations were united, with the usual political bickering and squabbling was put into cold storage.

The last Apollo's, 14-17, were most scientifically oriented. I was actually in Mission Control at NASA for the last of them; previously the BBC policy – of which I thoroughly approved – had been to leave me in London while the rest of the team went to the United States, so that I could hold the fort in case of any breakdown in communications. (This applied after No. 13; remember Moore's Law.)

In November 1970 Neil Armstrong was in London, and joined me in a television programme. We were not in the Television Centre, but in the foyer of a London hotel, where Neil was scheduled to appear first on one of the BBC's prestigious news programmes and then with me. During the first broadcast he was bombarded with all the usual snide questions: 'Does it justify the cost?' 'What about feeding the starving millions?' 'Can you really claim that it has been worthwhile?' and so on. He fended them tactfully, but as soon as he crossed the floor to join me for *The Sky at Night* he sank into an armchair and said, thankfully: 'Now we can talk some science.' We did.

Then there was the Lunar Landing Simulator... This was more or less on the lines of an RAF Link trainer, suitably modified, so that the 'astronaut' had to work everything out, conserving his fuel until he was ready to touch down at zero velocity. We were joined by one of the Moon-men (just this time, I won't give his name!) and we tried it out. I went first, and more by luck then judgement made a safe landing. Then came the astronaut. At the end of his approach the simulator flashed a sign: THERE WERE

NO SURVIVORS. LUCKY IT WAS ONLY A PRACTICE. As he had actually done it on the real Moon and I hadn't my faith in the simulator was somewhat weakened.

Much later, in 1982, I talked in Houston to Eugene Cernan, commander of Apollo 17 and the last Man (so far) to walk on the Moon. We intended to record a two-minute interview for *The Sky at Night* anniversary (our 25th year), but as soon as we began I knew that we had something really special. I glanced at Pieter Morpugo, who was directing as well as producing; he gave a quick nod, and so we went on for twenty minutes. It made a complete programme, and all we had to do was top and tail it. I asked him what was his greatest memory. 'The Earth,' he said. 'Looking up into the black sky and seeing my home world, a quarter of a million miles away.'

On the whole, our pre-Apollo ideas about the Moon were not so very wide of the mark. There is virtually no atmosphere; there is no life, and never has been any; the seas were never water-filled, though they were once oceans of lava. Nothing moves, the flags planted by the astronauts do not flutter, and will remain where they are until somebody collects them. So, too, will the Lunar Rovers or Moon-cars used by the last three expeditions. They remain on the lunar surface, and we know exactly where they are, and there is nothing to damage them, so that they remain fully functional. In the future a new Moon-man will go up to them, put in a battery, and drive off to the nearest lunar museum.

When that will be I do not know, but to quote Neil Armstrong when he talked to me some time ago: 'In some ways the Moon is more hospitable than the Antarctic. There are no storms, no snow, no high winds, no unpredictable weather; as for the gravity – well, the Moon's a very pleasant place to work in; better than the Earth, I think.'

I still see the Moon-men from time to time, and I would hate to lose touch with them. When television snared me for *This is Your Life* David Scott sent over a specially made tape, which I regarded as a great honour. At 'Expo 1989', at Brisbane in Queensland, Jim Irwin and I were the first speakers, Jim was, of course, David Scott's companion in the foothills of the Lunar Apennines. At the end of the Expo function, Jim and I were given wristwatches. Jim's watch strap was too big for him and mine was too small for me, so that today my strap has several of Jim's links in it. Sadly, Jim is no longer with us, he died some years ago now. A delightful man, quite unlike the swashbuckling hero of fiction, but a hero by any standards.

As recently as 2002 I was given the BAFTA Award for reasons I can't explain, and was deeply honoured when Buzz Aldrin flew over specially to make the presentation. It was a great occasion, and at the risk if seeming big-headed I cannot resist quoting what he said:

> 'I am pleased to say that this special award from the Academy tonight is being presented to my good friend the presenter of *Sky at Night* on BBC1, Sir Patrick Moore.
>
> '*The Sky at Night* began somewhat on the whim of a BBC executive, called Paul Johnstone, who wanted to make a programme or two to be called *Star Map*. 44 years later Patrick is still at it and still has no contract with the BBC. He says it's just a gentleman's agreement.
>
> 'Patrick is a self-made and self-motivated man. At the age of 11 he was elected the youngest member of the British Astronomical Association. Fifty years later to the day he became its President.
>
> 'On his 1908 typewriter he has written over sixty books on astronomy. And yet he still claims to be an amateur, although one of the few to be honoured with the CBE and OBE, seven doctorates and a Knighthood.
>
> 'Patrick's special subject has always been the Moon, he has named some features on it and has provided maps. In 1959 he was able to bring viewers the first direct pictures of the far side of the Moon. Incidentally it was his lunar charts that the Russians used to correlate this new information.

And again, it was some of Patrick's records that NASA used in the early Apollo landings.

'Not only has this man met every single lunar astronaut, he will modestly tell you that he has also met both the first man in space and the first man to fly in an aeroplane, Orville Wright, as well as the author H. G. Wells and the great physicist Albert Einstein.

'Here is just a glimpse of what he has been up to over the last years on television.' (*There followed a sequence of film clips*). 'I am pleased and thrilled to offer this BAFTA Award to Sir Patrick Moore.'

I replied:

'Thank you so much. I must admit that I am overwhelmed; there are so many people here who have done so much more than I have. After all, I have merely done some commenting.

'I suppose I did play a minor part in mapping the Moon, but I have a sort of feeling that Buzz knows rather more about the Moon than I do! And I do feel honoured that he is here to present the Award; it is great of him. If Buzz's trip in Apollo 11 had gone wrong, the Space Age would not be where it is now. He was the first; all honour to him. He name will live for all time; mine certainly won't.

'All I can say is that I don't think for one moment that I deserve this Award, but I am more than grateful. Thank you indeed; it has been a day I will never forget.'

Yes, I have lived through those pioneer days. And nobody can doubt that Neil Armstrong and Buzz Aldrin were ideally cast as the first two men on the Moon.

9 How's That?

There are some people who are natural athletes. Others are not, but by dint of hard work and application can attain a reasonable standard. Yet others are hopelessly unathletic, and can never be anything else. I have to admit that I belong to the latter category. I am clumsy and un-coordinated: it has been said that I give every impression of having been somewhat hastily constructed, and I would not demur. Moreover, before the age of fifteen I was not fit enough to play any games at all, so that through no fault of my own I was a late starter. Fate was not on my side.

Despite this, I have played a great deal of cricket, and am fanatical about it. Please do not misunderstand me. I have never been anything but a village greener. As I have the honour of being a Lord's Taverner I have taken the field with Test players and other deities, but I am under no delusions about my own standard. Moreover, I am hardly versatile.

I have taken a great many wickets with my curious leg-breaks, delivered at medium pace off a long, kangaroo-hop run and a cartwheel action; my model was the great Kent and England bowler D. V. P. Wright. I take fourteen paces – no more, no less – and my 'fast' one is, I suppose, fast-medium. I would not claim to be reliable, and I did ask one of my captains to refrain from crossing himself so obviously every time I came on, but I did spin hard, and it was said that my googly and my fast one were difficult to detect because of my weird action, which was once likened to that of a wallaby doing a barn dance. It was said of Ironmonger, the Australian left-arm bowler, that in the field he couldn't stop a tram, and I come into the same category; I am a poor catch, a poor throw, a poor stop, and I am as slow as a house. As for my batting… well, I am in the Hollies-Bowes–C. S. Marriott class. I have never batted away from No. 11, and I have taken many more wickets than I have made runs. I have two strokes, a cow-shot to leg and a desperate forward swat, which I play in strict rotation. I well remember that once during the 1950s, when I was captain of the local team and had reached the last day of August without actually having made a run, the opposition bowler very sportingly decided to present me with a slow full-toss to get me off the mark. Unfortunately he sent it down straight.

Despite a damaged left knee, which was always a nuisance, I was able to play most seasons between 1947 and 1999. My seasonal ambition was to take a hundred wickets and make a hundred runs. The wickets I usually managed up to around 1990; the runs, never. As a life member of Sussex I was able to go to Hove and practise with the professionals; unlike most amateurs I was always welcome, because I never bothered to take a bat with me, and we leg-spinners are an endangered species. Once, however, someone – it may have been George Cox – persuaded me to put on pads and face a few balls. After around five minutes he nodded. 'Yes,' he said. 'I see what you mean!'

Top-class cricketers are, as a rule, splendid people, and I have known a good many. I recall a match of long ago, when I was in uniform and was in an RAF team. Our keeper said to me: 'Kid, you're a shocker to keep to. You have this weird run, you have a lot of turn, and you're difficult to read.' I said, apologetically: 'Sorry, Leslie!' As a wicket-keeper-batsman, nobody has ever surpassed Leslie Ames.

When I came to Selsey, in 1967, one of my first acts was to contact the Cricket Club, where my reception was very different from that I had had at Armagh. For a time I was President, but then we lost our secretary, who left the area, whereupon I resigned the Presidency and was at once elected secretary. We had our ground on the recreation field, at the back of the fire station, but we had no pavilion, and had to change under primitive conditions (occasionally, under the hedge). We decided to build a pavilion, and we did. There were three Test men who voluntarily came over to support us: John Snow, John Edrich and Geoffrey Boycott. Rather later, Geoff paid me the compliment of asking me to be a speaker at a dinner given for him in honour of his hundredth century in the first-class game. I doubt if anyone knows more about cricket than he does, and I have a tremendous regard for him in every way.

Geoff is, I think, typical of my second type of athlete. He would not claim to have the natural talent of, say, Denis Compton, but by sheer perseverance and brain-power he made himself into the best opening batsman in England and one of the best of all time. Also, do not forget that he was an extremely useful change bowler.

Who would you nominate as the best of all left-arm spinners? I think most people, including me, would say 'Derek Underwood'. Universally liked as well as admired, he generally came out on top – but not always. Some years ago the Taverners repaired to Sevenoaks, in Kent, to play a match against the Blind School. Derek and I were the bowlers, and we had to follow certain regulations; for example, the outsize ball, with lead weights inside to make a noise, has to bounce twice before it reaches the batsman. We didn't know the rules well enough, and we lost. As I remember saying on the return journey, Derek could skittle the Australians and the West Indians, but had to admit defeat when faced with the Blind School!

One interesting match in which I was involved was played in Alderney, in the Channel Islands, soon after the island's liberation at the end of the war. The Germans had turned it into a concentration camp, and many of the fortifications still remain. The cricket ground is delightful, near the top, of a cliff, and as soon as the Germans had been removed play was resumed. I was invited to play for the Visitors against the Islanders, so I flew over in the 'little yellow plane' and prepared to do battle. We batted first, and when I went in we were about 100 for 9. The first ball I received was a rank long-hop; I swung at it, connected, and sent the ball over the cliff into the sea for six. The next ball was an equally rank long-hop outside the off stump. Again I swiped; again, by sheer luck I connected – over the cliff and into the sea for six. That was the end of the match. There were no more balls, and all we could do was to adjourn to the bar in the Harbour Lights pub. Even today, in Alderney, that truncated match is still remembered!

Odd things can happen in my class of cricket. Once, in Sussex, I bowled a short ball which was hit into the outfield, where it struck a goat and was caught by long leg. Was it out? The umpire thought so, and presented me with an undeserved wicket. But does a goat come into the same category as an umpire; making the ball dead? Arguments raged up and down the pub after the match, and there was no general agreement, so we referred the whole matter to the MCC – who wrote back a rather stuffy letter pointing out that the goat shouldn't have been there. On another occasion our fast bowler was hit for a twelve, because the ball vanished into a rabbit-hole in the outfield and the fielder omitted to call 'Lost ball'. By the time he had located it the batsmen had run eight, and a wild return to the keeper resulted in an extra four overthrows.

Near Guildford, in Surrey, we used to play a match on the last Sunday in February in aid of the local children's home. It was a matter of honour to go ahead, whatever the weather. One year I recall Ken Barrington and (I think) John Inverarity trying to bat, with Jim Laker at one end and me at the other doing our best to bowl off two paces on pack ice. Even Jim couldn't keep a length on that one! Quarrels on the cricket field are rare in my experience, but I did once have a dispute with the local Vicar, who dropped two catches in the slips off me in the same over. I bought him a drink in a marked manner, and said acidly that although Christ might forgive him, I wouldn't.

I have never taken all ten wickets in a good Club game, but on several occasions I have come close to it, and once I feel that I was decidedly unlucky. We faced a strong team; I had nine wickets, and in came the opposing No. 11, who was admittedly a rabbit (though a very fine fast bowler). I sent down a leg-break. How it missed the wicket I know not, but it did, and the batsman ran; the non-striker did his best, but had no hope, and was run out by yards, leaving me with 9 for 45. That was one of the few moments in my life when I seriously considered committing suicide!

But I have had my moment of glory, and I am conceited enough to describe it here.

Cast your mind back to 1947, that blazing hot summer when the South Africans were here, and political trouble-makers had yet to emerge from under their flat stones. All through the summer Edrich and Compton butchered the South African bowling, and Wright tore through their batting. I was playing much more modestly in Sussex – I was, indeed, team captain – and in early June we found ourselves up against a very strong side from Tunbridge Wells. Normally, they would be too much for us. We were reasonably good with the bat – I was the only total rabbit – but we had precisely four bowlers; one fast, one who believed that he was fast, one slow left-armer, and me. Actually I was having a good season, and on the previous weekend had returned an analysis of 7 for 28 – plus a dropped catch by the keeper, for which I mentally consigned him to the deepest recesses of hell. With the bat I was not so successful. My most recent scores had been 0, 0, 0 not out, 0, 0, 0 and 0.

It was a glorious day (as were most days that summer), and I won the toss. I had no hesitation in deciding to bat. (Remember W. G.'s advice? 'If you win the toss, bat. If you have any doubts about it think for five minutes – and then bat.') We were at full strength, and our Nos. 1, 2, 4 and 6 were of really excellent Club standard, so I hoped for great things. They didn't happen. Tony and Bill, our openers, put on about ten, but then the rot started. The main opposition bowler was a solicitor by profession, and he was decidedly fast; there was also an off-spinner who never seemed to bowl a bad ball. Tony was stumped, and Bill made a wild and quite uncharacteristic lunge, with the inevitable result. Wickets tumbled. Suddenly, with a feeling of horror, I realized that we were 27 for 9, with Adrian, our No. 6, still in but not looking at all happy. He was always a defensive bat, but with only me to come it was clear that he would have to hit out and shield me.

I trudged out to the wicket, all eyes on my forlorn figure (we had a good crowd of watchers). My batting was only too well known, and the fielders licked their lips in anticipation. The fast man was on, and as I took guard I fully expected to be skittled. I was then aged 24, but with the bat I had none of the brashness of youth.

The bowler thundered up, and unleashed a lightning-bolt. I failed to see it, but there was a satisfying crack as my bat connected, and a ripple of applause as the ball rocketed to the boundary. Well, at least I had opened my account. Next ball – the same result. Again I lost it completely, and again, it soared over slip's head to the boundary. Eight runs – almost a career best. I began to feel that the gods were on my side.

Adrian faced up to the next over. He edged one, and called. It was safe enough, and we ran a single, which took some time, because my speed between the wickets has been compared with that of a particularly leisurely cortège. Adrian wanted two, but no luck, and had five balls to face. Three of them went to the boundary, and one of them I actually saw, though to pretend it went where I intended would be what has often been called terminological in exactitude. The next was a simple caught and bowled, which was grassed. The slow man muttered something under his breath. Off the last ball I edged a catch to second slip, who dropped it, and we ran a single, giving me the strike once more. Adrian's face was a study.

So it went on. I played my two strokes in rotation, as usual, but by a series of miracles I kept connecting. At the interval I was 36 not out – a score which I had never previously approached. Adrian was marooned on 8.

During the lunch break there was considerable comment. One of our batsmen (No. 4, actually) was the Vicar, who informed me that I had so far survived purely by divine intervention. I urged him to keep on praying.

I enjoyed lunch. Then, back to the field. I clouted the first delivery from the mast bowler for six; I aimed it over the bowler's head, but in fact it went over the wicket-keeper's head; what part of my bat connected with the ball I know not. In almost no time at all I realized that I was on 48. Adrian was by now totally incredulous, but I was in my element. I went for a big hit; I edged the ball to cover – and once more it was put down; we took a single. 49 not out. Could it really be possible that I would complete my half-century?

I did – off another dropped catch, this time in the deep (and I don't believe that our opponents had grassed a single catch in all their matches to date). As we completed our two runs there was a roar of applause that could probably have been heard in Bognor Regis. In the fifties I was dropped again, and finally, when I had made 63, the end came. Occasional bowlers had entered the fray by now, and one of them delivered a very gentle long-hop. I hit it, but in the process I also hit my wicket, and that was that. Moore, ht wkt, b Ferris, 63. The next highest score was Adrian's modest 9 not out.

Well, I thought, at last I will achieve a lifelong ambition of making a hundred runs in a season; we had about twenty matches to play, and I ought to have a good chance. I had visions of a Test call. Alas – it was not to be. We did win that particular match, though my own contribution was an undistinguished 1 or 18. For the next fixture I was urged to put myself away from No. 11, but

I was too wary for that. When I was castled first ball, I was not disconsolate, but at the end of the season my total number of runs had risen to no more than 68. You can work out my average from that, more or less, if you are unkind enough to do so.

I wish I had a film record of that one innings. I will not forget it, and neither will our opponents, even though it happened well over half a century ago now. The local newspaper commented that I had luck enough for two whole teams, but the final verdict was given later by the Vicar, when we retired to the pub after stumps were drawn. The pub overlooked the sea, and the Reverend pointed to a distant sandbank. 'Care to walk over to that?' he asked. 'If anyone is capable of walking on water, it must be you!'

I didn't try. Miracles don't happen twice.

For the sake of completeness I may as well say something about other games, but it will not take long, because, as I have said, I am no athlete. Golf is great fun, and I did once win a cup. It was in a Pro-Am competition at Southampton in 1975, or thereabouts. The amateurs were paired with professionals, and some of the amateurs were really good; Cliff Michelmore was playing, and he is an expert (he was also a cricketer of County standard; I envy him). All the pairs recorded scores of between 64 and 75, except me. My drive off the third tee was superb, and ended up on the green – but of the seventeenth. My final score of 148 was so remarkable that the judges decided to give me a special award for introducing an entirely new standard to the competition, with the result that I won a cup and, on this occasion, Cliff just missed out.

We were again involved some time later. A new, probably illegal, club had been sent into the BBC, and was the subject of a television programme. The idea was that I would go first; with a normal club. I would miss the hole, and with the illegal club I would hole out. Cliff would then follow, holing out with the regulation club and missing with the other. It didn't work out quite like that. I could not hole out, and Cliff couldn't miss. We finally got the shot we wanted in Take 15!

My last golf match, before I was forced to give up, was against an old friend, Michael Foxell, whose standard was comparable with mine. This was a needle match; a bottle of whisky was at stake. We came to the last hole all square. This was a par 4, and Michael holed to in a chanceless 17, which was far too good for me. Actually I wasn't too bad a putter, but for the rest – well, silence is golden.

I wasn't too bad at table tennis, but the only game in which I really could claim to be fairly good was chess, which has always been a hobby of mine. I learned it when I was very young, and some people were kind enough to say that if I had concentrated on it I might have made my mark, but of course I never did, there was never enough time. I played for Sussex, and I still play some correspondence games. One thing that beats me is this new algebraic notation which has widely replaced the old classic form. If I could meet the man who invented it, I would take sadistic pleasure in peeling him like a banana.

During the 1950s, when I was a serious player, I had three games with a boy aged six. Apart from his chess, he was a perfectly normal lad – but on the chessboard he was not only out of my street, but out of my parish. He could have given me a rook, and possibly a knight as well. I could never have beaten him, and I wondered whether we had a new Bobby Fischer in the making. We will never know, because he was killed in a motor accident a year later. I was very sad; he was a thoroughly decent boy, and I am sure he would have made his mark.

I did not distinguish myself when the New Scotland Yard chess team challenged the Lord's Taverners (!) to a charity chess match. With the Press in attendance, we assembled in New Scotland Yard; ten of them, ten of us. As Board No. 1, with a high ranking, I was the star of the Taverners, and expected to win comfortably. In fact, not only did I lose, but I was first to lose. I have never played so badly in my life, and there was not the slightest excuse. I then had a friendly game with the same opponent, and I was far too good for him, but it was too late then.

On the credit side, I did play a draw with a Grand Master, Bronstein, in a charity event in Hastings. The game was played under championship conditions, and a draw was agreed after 34 moves, but it wasn't really fair, because only half-way through did he realize that I am fairly strong. Had he known that from the outset, a draw would have been beyond me.

I am rusty now, but I enjoy correspondence chess, and one of my regular opponents is Mike Fox, secretary of the British Chess Federation. Our games are, to put it mildly, unusual. Of course, he is completely out of my class, and in a face to face game under normal conditions I would have as much chance against him as a pork butcher would have of making a living in Tel Aviv, but I do enjoy our battles, and occasionally I steal a game – as I did here:

White: Moore	*Black:* Fox
(Demented Opening)	
1 Kt–QB3	P–QKt3
2 P–K3	B–Kt2
3 P–Q3	P–KB4
4 Kt–B3	Kt–KB3
5 P–KB4	P–KKt3
6 Kt–KB4	B–Kt2
7 B–K2	P–K4
8 Kt–B3	P–Q3
9 Q–Q	Qkt–Q2
10 Kt–K	Kt5Q–K2
11 B–B3	P–Q4
12 Kt–Q	Kt5P–KR3
13 Kt x Pch	K–Q1
14 Kt x R	P x Kt
15 Kt x P	P x Kt
16 P–QB4	Q–B4
17 R–Kt1	P–B5

18	P–QR3	P–K5
19	P x KP	QP x P
20	B–Kt4	Q x P
21	B–K2	Q B3
22	P–QR4	KT–Q4
23	B–QKt5	Q–B4
24	P x P	P x P
25	B x P	B–B1
26	Q–Kt4	B–B1
27	QR–B1	Q–Q3
28	B–KT5 ch	Kt(Q4)–B3
29	KR–Q1	Q–K2
30	Q–R4	P–K6
31	B x P	R–R1
32	Q–Q4	Kt–K5
33	Q x P ch	Resigns

Needless to say, in the return match he annihilated me.

I wish Mike had faced that six-year-old boy from Bognor. He would have won, of course, but the lad would have given him a real fight.

10 Tally-ho!

I never get involved in unpleasant arguments unless there is absolutely no alternative, if only because they are bound to concern unpleasant people. Generally speaking I manage to steer clear, but there is one major exception: a battle which has been going on for many years and is not over yet, though the end is nigh. This is about hunting animals with dogs.

I suppose that so far as I am concerned it began one day in 1938, when I was fifteen and just emerging into the real world (up to then I had been more or less isolated, through no fault of my own or anyone else). With two friends I went down Worsted Lane, where my home lay, and we wandered into the fields adjoining the local farm. We had guns with us; I can't imagine why, because a gun was the last thing I would normally carry, and I certainly didn't own one. Naturally the gun I had on this occasion was little more than a toy, but it *could* have inflicted a very nasty wound.

Rabbits were running around, and one of my friends pointed. 'Have a pot at that one.' So without thinking I pulled out the gun, aimed at the rabbit, and fired. Of course I missed it by a mile, because I am the worst shot in three continents, and the rabbit was never in the slightest danger. But as it scuttled off to its hole, a sudden thought struck me. If I had been more skilful I would have hit that little animal, causing it intense pain and possibly taking its life – and I had tried to do this *just for fun.*

I felt utterly ashamed of myself, and needless to say I never again shot at an animal; even now, after a lapse of over sixty years, I am thankful that I missed (I hope that my intended victim had a long and happy life!). Ever since then I have had deep contempt for people who go out to kill merely to amuse themselves. And of all these the most despicable are the foxhunters, the stag-hunters and their kind.

Please do not misunderstand me. I am no woolly-minded, ultra-Green vegetarian, and I know that if we did not eat animals they would end up by eating us, and because I do eat meat and fish I can be regarded as a hypocrite, even if I do try to kid myself that a pork chop grows on a tree. (An uncle of mine once turned vegetarian on principle. This lasted until Christmas, when he protested volubly that a goose was a fruit.) Nature is cruel, and we are part of it, but there is a world of difference between killing for food or protection and killing for nothing more than mere pleasure. Moreover, if you shoot an animal or have it humanely slaughtered, you do it as quickly and painlessly as you can, whereas fox-hunting, stag hunting, otter hunting and hare coursing involve prolonged torture – there is no other term for it.

What attracts people to deliberate cruelty? I saw enough of it in my formative years (late teens and early twenties) to sicken me. After the rabbit episode I became actively involved in the anti-hunting campaign. My parents agreed with me, and so did my friends, so that there was no friction there.

There are two main aspects, as I realized when I began to think seriously about the problem. First there is the cruelty to the animal, which is frankly barbarous, but we must also consider the effects upon people who hunt, particularly youngsters. I can remember what I said during a television interview quite recently:

'Look at it this way. A boy (or girl) is taken to a hunt. He puts on fancy dress and gets on a horse. He and the rest then gallop along, whooping like dervishes, until the dogs pick up the scent of a fox. The fox is then chased until its lungs are bursting, and is eventually caught. You watch, cheering, as it is torn to pieces. The huntsman then picks up a breeding stump and wipes your face with it, a ritual known as "blooding". That ends the fun, for the moment.'

Well – is that the right way to turn out a decent, civilised adult? I subsequently argued with a most unpleasant Tory woman M. P. who maintained that hunting was part of the British way of life. (*En passant*, the hunting lobby has claimed that blooding no longer takes place – but of course it does.)

Hunting is cloaked in hypocrisy, and I have even heard it said that the fox enjoys the chase! Also, it is said that hunting keeps the number of foxes down; on the contrary it keeps them up, and in case of shortage the huntsmen ferry cubs from one area to another. Of course alternative methods of fox control, such as gassing and shooting, can cause pain, but the cruelty here unintentional, not deliberate. As for 'cubbing' – so far as I am concerned, anyone who takes part in this peculiarly loathsome 'sport' must come from another planet.

Note, too, that the spurious arguments advanced by the foxhunters cannot possibly apply to stags, and indeed all forms of blood sports come into the same category; this includes dogfighting, which is illegal in Britain but is no worse than stag-hunting. We have at least got rid of otter hunting; how anyone could take part in that pursuit I know not.

Viciousness is another characteristic of hunting folk. I well remember Bob Churchward – Captain Robert Churchward, M. C. – a dedicated huntsman who changed sides, and 'blew the whistle'. To say that he was unpopular with his erstwhile friends is to put it mildly, and he was threatened physically. So, on one occasion, was I; it was in 1966, when I lived at East Grinstead. I had persuaded two of the farmers to close their fields against the hunt, and I was informed that two members of the local pack were going to come and beat me up. Rather to my regret, someone warned them that I was rather good at that sort of thing (I had been extremely well trained), and they failed to put in an appearance!

The League Against Cruel Sports is doing all it can, but of course the key organization ought to be the RSPCA, and here there have been some curious episodes. When I first joined, in the 1950s I was staggered to find that not only did the Society support hunting as a method of fox control, but was crammed with people who actually hunted, and the members of the Council were distinctly unprepossessing. The Chairman was the ineffective Colonel Lockwood; there was a Mrs. Tait, who said openly that 'the RSPCA is pro-hunting'; the Vice-Chairman was Dr. Rattray, whom I regarded as a very nasty piece of work, and there was an odious clergyman named Snell, often referred to as 'Slimy' Snell. It was noticeable that most of the hunting gang clung on to out-of-date military rank. Lockwood had at least been a proper colonel, but I well remember one man whose name was something like Nanki-Poo; he called himself a captain, but my guess is that he would have run a mile if he had encountered an angry German.

Special Extraordinary Meetings on the hunting issue were called, mainly by Miss Gwen Barter, Howard Johnson (MP for Kemp Town, Brighton – a Conservative) and me. By various dodges the Council managed to block any anti-hunting motion, and Gwen Barter and Howard Johnson were expelled from the Society. I would no doubt have been expelled too, had I not resigned in disgust. I did not re-join until 2001, and by then the situation was very different.

I am not sure how the change came about, because I was out of the RSPCA picture, but during the 1980s the hunting members were removed, decency took over, and the Society committed itself to a policy of a hunting ban, which brings me on to 'modern' times.

Politics came into the arguments, and the hunting lobby claimed that all 'antis' were state-educated townies, belonging to what was called the working class, and who had no appreciation of country life. In that case, I wonder where people such as me fit into the equation? I am country born and bred; only ill-health kept me away from Public School, and my family is fairly well represented in the pages of *Debrett*. But it is true that in general the Conservative Party supports hunting, while the Labour Party does not (I have no idea where the Liberal Democrats stand on the issue, but of course they never have original ideas about anything). In a recent debate I saw Mr. William Hague on television making a passionate defence of hunting; watching him, I wondered how on earth the Tory Party had selected such a twerp as their leader (as least they had enough sense to ditch him before long, and hand over the leadership to the amorphous Mr. Smith). However, not all Conservatives are tarred with the same brush – Ann Widdecombe is doing all she can to outlaw hunting, for example – and there are some renegades on the political Left.

When Labour came to power with a crushing Parliamentary majority, many people – including me – thought that anti-hunting legislation was on the cards. However, a very noisy and well-organized opposition managed to hold things up. Tony Blair, as Prime Minister, promised and indeed gave a free vote; the House of Commons was whole-heartedly anti-hunting, but the Conservatives were able to strangle the Bill by 'talking it out'. When it was re-introduced, in the next session, the vote against hunting was even more emphatic, and the Bill went to the House of Lords, where it was – predictably – thrown out. Again it was introduced in the Commons; again the vote was overwhelming, and the Lords could not block it again, because it could be forced through by using the Parliament Act. Yet to date (January 2003) this has not been done, and the Government is showing signs of trying to fudge the issue. By the time that this book is published the situation may well have been clarified. At least Scotland has taken the lead, and hunting there is now illegal.

One man who shows up in a very bad light is Mr. George Carey, the former Archbishop of Canterbury. He is Vice-Patron of the RSPCA, but has refused to give his personal support to the

Society's anti-hunting stance. He has been asked for a straight 'Yes' or 'No' answer as to whether he disapproves of hunting, but he will never give an answer; all he does is to send non-committal replies via his secretaries, and he will not reply personally to MPs, members of the general public, or even the RSPCA Council. I can see only two possible reasons for his reticence. Either he does not disapprove of hunting, in which case he is a hypocrite for remaining as an official of the RSPCA, or else he does disapprove and is afraid to say so, in which case he is a moral coward. The Church might lose revenue from wealthy members of the unsavoury 'Countryside Alliance', so this may well be a factor, but by refusing to come out of his closet Mr. Carey is damaging not only himself but also the Church (which matters to many people) and the RSPCA (which matters very much indeed).

His predecessor as Archbishop, Mr. Runcie, was equally adept at wriggling; when I wrote to him commenting on the obscenity of Boxing Day fox-hunts, he replied that since December 26 was not officially a holy day, it didn't count! Every Archbishop seems to be worse than the last. Mr. Carey was succeeded by Mr. Rowan Williams. I wrote to him on the hunting issue. Unlike Mr. Carey he did have the decency to reply, but so far (January 2003) he has not given his personal support to the RSPCA campaign. I hope that he will eventually do so, if only because reconciling Christianity with animal cruelty seems to me rather difficult.

No decent-thinking person will doubt that the end of hunting will remove something squalid and cowardly from our way of life. Hunting must be outlawed soon; when it is, I will be proud to have played a part, albeit a very minor one, in its demise.

11 Mars Hill

One of my favourite places is Mars Hill, in Arizona, just outside the town of Flagstaff. It was here that Percival Lowell established a major observatory, and equipped it with one of the best telescopes available at the time. His main aim was to use the telescope to search for life on Mars.

Is there life on other worlds? I have lost count of the number of times I have been asked that question. I cannot give a definite answer; all I can say is that I find it impossible to believe that we are alone. Our star-system or Galaxy contains around 100,000 million suns, and we can see at least a thousand mission galaxies; we are now sure that many stars have planetary systems of their own, and the number of 'Earths' must be staggeringly great. There is no reason why there should not be other men, very possibly similar to ourselves. As yet there is no proof, but the clue may come from Mars, though not in the way that Percival Lowell expected.

(*En Passant*, note that I say 'other men.' This is no doubt Politically Incorrect, and will anger strident feminists, but I must point out that the correct name for the human species is *Homo sapiens*, and to use anything else is a confession of ignorance.)

Of all the planets in our Solar System, only the Earth is suited to intelligent life, and conditions elsewhere are not promising. Things did not seem to be quite so bleak half a century ago, when it was still thought that there were wide vegetation tracts on Mars even if there were no Martians. Lowell had used his telescope to draw 'canals' crossing the Martian deserts, and claimed that they indicated the presence of a highly advanced civilisation capable of carrying water from the polar ice-caps through to the warmer regions at the equator.

During my Moon-mapping days I spent a long time at Flagstaff, and I came to know Mars Hill very well; I did not meet Lowell – he died in 1916 – but I know people who did, and he was above all an enthusiast. His telescope collected its light by using a lens 24 inches across, and even today there are not many refractors larger than that. During my first spell at the Observatory Mars was high in the sky, and I had my first view of it through Lowell's telescope. With me was Charles ('Chick') Capen, whose experience as a planetary observer was second to none. We turned the telescope toward Mars, and I settled myself at the eyepiece.

I admit that I was excited. Would I see canals, as Lowell believed he had done? There were the polar ices, the ochre 'deserts' and the dark patches still thought to be due to plants, but canals – no. They simply were not there, even though conditions were good, with a rock-steady telescopic image, and I was using a high magnification. At last I looked round. 'I can't see them.' Chick chuckled. 'Neither can I. They aren't there.'

I was delighted that I had failed. The canals were nothing more than tricks of the eye; it is only too easy to 'see' what you half expect to see, and this is where Lowell and his followers went wrong. I would have been very annoyed with myself if I had fallen into the same trap.

All the same, there seemed no reason why life of some kind might not exist, and in a *Sky at Night* programme in 1950 (before the first spacecraft to Mars were launched) I worked out an interesting experiment. I enlisted the aid of a friend, Francis Jackson, a skilful amateur astronomer who was a microbiologist by profession, and we decided to make what was to all intents and purposes a Martian laboratory.

We needed to know what conditions on the planet were like, and we used the latest information available. We knew the gravitational force; it was one-third that of the Earth, so that if you weigh 12 stones on Earth you will tip the scales at only a modest 4 stones on Mars (actually you will have to use a spring weighing machine). This was something which we could not simulate, so we had to hope that the increased 'weight' of our samples would make no essential difference. We knew the length of the Martian day, just over half an hour longer than ours (to be precise, 24 hours, 37 minutes, 22.6 seconds). The length of the Martian 'year' – 687 Earth days, or 669 Martian days – did not come into it, as our experiments were to be short-term. At noon on the equator, in summer, the temperature could rise to over 40 degrees Fahrenheit, but the thin atmosphere was very poor at retaining warmth and the night temperature sank to around 150 degrees. As I remember saying, and campers on Mars should remember to take plenty of warm clothing with them. (Incidentally, I am not fond of camping. Once I was persuaded to go, with two friends, and made for Cornwall. One night I was cook; when I passed over a plate of fried sausages I realized that one sausage was not a sausage at all, but a slug. I never let on, so apparently it was quite tasty.)

There remained the problem of the Martian atmosphere, and this is where we went badly astray, though at the time we had no chance of doing any better. Careful studies carried out from Flagstaff and elsewhere indicated that the main constituent of the atmosphere was nitrogen, which makes up 78 per cent of the air we breathe on Earth, but the atmospheric density was low. The favoured value for the pressure at ground level was 85 millibars, in which case the ground pressure was the same as that in the Earth's air at a height of 52,000 feet.

Everest, our highest mountain, is less than 30,000 feet above sea level, and even there climbers find it difficult to breathe without using oxygen masks, so there was never any chance that future explorers would be able to go to Mars, step outside their space-craft and take deep, refreshing breaths in the mid-day cool. But lowlier forms of life are surprisingly hardy, and a nitrogen atmosphere of this kind would be quite suitable for organisms such as algae. We even anticipated lowly vegetation. Why should not the dark tracts on Mars be old sea-beds covered with, for example, lichen or moss?

There was in fact good reason to believe that this might be so. The dust storms would otherwise spread the reddish, dusty material everywhere, and hide the dark regions, so that the whole of Mars would be a uniform ochre hue. The Estonian astronomer Ernst Öpik, at one time a colleague of mine at Armagh, pointed out that plants could push the dusty stuff away and poke through quite happily.

One thing we knew we would not find was advanced life; not even insects could survive under Martian conditions. At a public lecture just before our television programme I stated that 'on Mars there may be lowly organisms, but nothing so advanced as a cabbage.' One local newspaper gave a rather inaccurate report: 'Astronomer claims that the nearest approach to Earth-type life on Mars is the growth of large cauliflowers.' I gave up.

Under Francis' direction we constructed our Martian laboratory, which took the form of a large airtight cabinet. We filled it with a Martian atmosphere – chiefly nitrogen, with traces of carbon dioxide and other gases which we might reasonably expect; we adjusted the pressure, and arranged for the alternation of temperature between day and night. We thought we knew a great deal about the Martian soil, which was a mixture of iron oxides – rust, if you like – and various rocky materials. Then we discussed what organisms we should try.

Cacti seemed to be good candidates. They can flourish in the most improbable places; they need very little water, and in any case we would not rule out the possibility that there might be a water source underground. Cacti are also tolerant of violent temperature fluctuations between day and night. So in our first programme we produced a cactus, and announced that we were about to subject it to Martian conditions.

The cactus didn't like it, and after a single Martian night it looked decidedly limp. When we saw it we all broke into uncontrollable laughter, but we decided that we really could not show it on the programme; I won't go into the details of why not – I will leave those to your imagination. So the cactus was quietly removed, and we turned to simpler life forms.

The results were interesting. Very primitive organisms such as algæ were uncomfortable after a couple of Martian nights, but some types of bacteria did better, and there were even indications that they might be able to multiply. In our second (and, as it proved, final) programme on the subject, we went so far as to claim that 'if our picture of the Martian surface and atmosphere is correct, there is a real possibility that some terrestrial bacteria could establish themselves there.'

We felt that we had made a real contribution to science. Our experiment was not the first of its kind, but it had been carried out under strictly controlled conditions, and had been supervised by an expert. We waited for the academic world to acclaim us.

Alas! pride goes before a fall. And within a year or so we found out that we had achieved absolutely nothing.

In November 1964 the Americans launched their second attempted Mars probe, Mariner 4 (its predecessor, Mariner 3, had taken off in the wrong direction, and had joined the ever increasing swarm of useless artificial planets). On 14 July 1965 Mariner 4 passed within 21,0000 miles of Mars, and sent back close-range pictures, showing that instead of being no more than gently undulating, as everyone had expected, the surface was cratered like that of the Moon. Worse, it established that the atmospheric ground pressure was nothing like as high as had been believed. To reach equivalent density above the Earth you would have to go up not 52,000 feet, but more than twenty miles. In fact, the situation on the Martian surface is comparable with what we could call a reasonably good laboratory vacuum, and to complete the picture the bulk of the tenuous atmosphere consists not of nitrogen, but of carbon dioxide.

So the work we had done was based on completely erroneous data. The sacrifice of that noble, willing cactus had been in vain.

So far as I was concerned, the next serious involvement with Mars was in 1971. The Americans had sent up a new probe, Mariner 9, which was by far the most ambitious to date, inasmuch as it was designed to go into orbit round Mars instead of merely passing by. (In case you are wondering about Mariners 5, 6, 7 and 8; 5 was sent past Venus, 6 and 7 flew by Mars, and 8 did not fly anywhere; its motor failed immediately after launch and it plumped down in the sea.) In the late summer of 1971 Mars was about as close to us as it can ever be – around 35,000,000 miles away – and it was well South of the celestial equator, so that as seen from England it was low over the horizon. Mariner 9 was scheduled to reach its target in November, and it duly did so. It began taking pictures as it circled Mars, and sent them back to Earth for analysis. Transmissions did not cease until the following October.

There was an immediate problem. Normally the thin Martian atmosphere is fairly clear, but there can be major dust-storms, and these tend to occur when the planet is at its closest to the Sun (its orbit is much less circular than ours). There were reasons to believe that such a storm might take place in the winter of 1971 (our winter, I mean, not that of Mars) and the obscuring dust might easily hamper the work of Mariner 9.

I was sitting in my Selsey study, doing no harm to anybody, when the phone rang. It was from one of the NASA team. 'Will you go over to South Africa, and use the Johannesburg telescope to monitor the Martian storms?' I did not demur. As I have said, Mars was in the southern part of the sky; the 26-inch Innes telescope at the Republic Observatory was well known to me (it was even larger than Lowell's), and I was happy to go. A couple of days later I was in Jo'burg. The astronomers there made me welcome, and there was no objection to my using the telescope, which at that moment was not busy on any official programme.

Mars was indeed in the grip of a dust storm. At first I could see little but a blank, reddish ochre disk, and there was nothing I could do except wait for the dust to subside. Eventually, after a few weeks, it did, and the tops of the giant Martian volcanoes poked out; slowly the atmosphere cleared, and Mariner 9 sent back amazing images of the volcanoes, the valleys, the craters and the features which are almost certainly dry riverbeds; long ago there must have been seas. I hope that my daily reports to NASA were of some use.

South Africa then was not the South Africa of today; it was quite safe to walk around at night provided that you kept out of certain areas (as indeed we have to do in Britain). I was a frequent visitor, and in fact made weekly broadcasts on South African radio. All this sort of thing came to an end when the Government changed, and Mr. And Mrs. Mandela took over. I no longer felt welcome.

Years later, in 1997, I went back to Johannesburg to do a broadcast for *The Sky at Night* Fortieth Anniversary programme. Naturally, I went to the Observatory, and to my dismay found an extraordinary state of affairs. The astronomers were not able to use the Innes telescope and the dome had been taken over for social functions such as weddings; the telescope had been damaged, and it was impossible for the astronomers to gain access to it.

I was in a better position. I managed to get into the dome (by bribing one of the security men), and took photographs. I then broadcast on South African television, explaining the situation. I did wonder how the authorities would react, but I am glad to say that the broadcast had the desired effect, though no doubt it was the spark to initiate action. Anyway, the telescope is again ready for systematic use, though I am not sure what the intentions are. I have not been back to South Africa, and now, I fear, I never will. I am also sorry that I never went to Rhodesia before we stabbed Ian Smith in the back and handed the country over to Robert Mugabe, with results which were entirely predictable.

On the other hand I have been back to Flagstaff, many times, and I have the happiest memories of it. It was here that I first met Clyde Tombaugh, the only living man to have discovered a new planet.

To put things in perspective, I must backtrack for a few moments. Five planets – Mercury, Venus, Mars, Jupiter and Saturn – have been known since ancient times. In 1781 William Herschel, a musician-turned-astronomer, found the next one, Uranus, which moves well beyond the path of Saturn and takes 84 years to go once round the Sun. It is so far away that it is only just visible with the naked eye, even though it is 30,000 miles in diameter. Slight irregularities in its motion led two mathematicians, Urbain Le Verrier in France and John Couch Adams in England, to conclude that Uranus was being pulled out of position by an unknown planet further away from the Sun. Independently, and without knowing about each other's work, they set out to find where the new planet must be, and in 1846 Johann Galle and Heinrich D'Arrest, at the Berlin Observatory, found it – almost exactly where expected. It was named Neptune, and proved to be almost the same size as Uranus.

Still things were not quite right, and around 1900 Percival Lowell made fresh calculations in the hope of tracking down yet another planet. He worked out a position and looked for it, using photography with the 24-inch refractor. He did not find it, and when he died suddenly, in 1916, the planet was still missing.

In 1929 Vesto Stipher, who had become director of the Lowell Observatory, decided to try again. He hired a young amateur, Clyde Tombaugh, who had sent in some excellent drawings of Mars, and set him to work, using a 13-inch telescope obtained specially for the purpose. Clyde's method was to take photographs of the same region of the sky separated by several days; the stars would not show any relative movement, but a planet would betray itself by its motion. After only a few months Clyde identified an object which moved in precisely the right way, not far from the position given by Lowell. It was indeed the awaited planet – named Pluto, partly because it is a world dimly lit by the Sun, and partly because the first two letters, PL, relate to Lowell. (The name was actually suggested by a young Oxford girl, Venetia Burney. I met her not so long ago; she is now Mrs. Phair.)

I first heard about the discovery of Pluto when I was seven. Little did I then think that fifty years later, the discoverer would invite me to collaborate in writing a book about it!

I met Clyde on my first visit to Flagstaff. He was a charming person, always ready to help others in any way he could and universally liked and admired. By then he had officially retired, but he still had an office in the Observatory – about the untidiest I have ever seen, though he seemed to have no difficulty in locating anything that he needed. His home was in Las Cruces, and it was there that he had set up a large reflecting telescope. Outwardly it looked like a heap of old iron, but it was in fact a superb precision instrument. One of the mirrors was still covered, when not in use, by a lid bearing the significant word COFFEE.

Clyde had often been asked to write the 'Pluto' book, but had always declined. When I was at the Observatory, he suggested that we might produce it together. I was deeply honoured – well, who wouldn't have been? – and in the fullness of time *Out of the Darkness: the Planet Pluto* was published. It was well received, though I must stress that I played a very minor rôle.

Incidentally, there are doubts about Pluto's status. It is much smaller and fainter than Lowell had expected (which is why he failed to find it) and it is even smaller than the Moon. It has a strange orbit, which sometimes brings it closer into the Sun than Neptune can ever be; its 'year' is 248 times as long as ours.

In 1980, fifty years after the discovery of Pluto, a symposium was held at Las Cruces, with Clyde as the guest of honour. I was delighted to be there, and it was a splendid gathering. During it, an asteroid was officially named 'Tombaugh'. In his speech, Clyde commented that he now had a piece of real estate that nobody could touch, even if it is more than two hundred million miles away and is a chunk of material less than ten miles across!

Recently, other small bodies have been found in that remote part of the Solar System, and it has been suggested that Pluto is merely the largest member of this swarm. Also, it is so puny, and has so feeble a gravitational pull that it could not possibly cause any measurable disturbances in the movements of giant planets such as Uranus and Neptune – and so either Lowell's accurate prediction was a sheer fluke (which I find hard to believe) or else there really is another large world out there. If so, we may hope to find it one day. Asteroids are dwarf worlds, most of which move round the Sun between the orbits of Mars and Jupiter; only one (Vesta) is ever visible with the naked eye. Asteroid Tombaugh is No. 1604. (Asteroid Moore, named after me years later, is No. 2602. I can just about see it with the 15-inch telescope in my observatory at Selsey.)

A tremendous amount of work is done at Flagstaff, and Lowell himself will never be forgotten. The one thing he was not, unfortunately, was a good observer. He was completely wrong about Mars, and he also drew streaky markings on Venus, though he did not suggest that they might be artificial. As soon as I had the chance I turned the great 24-inch refractor toward Venus, but I could see little apart from the characteristic phase: the clouds in Venus' atmosphere never clear away, and virtually all our detailed knowledge about conditions there has been obtained by space-craft. Previously, it had been thought that Venus might be a rather pleasant place, with wide oceans and luxuriant vegetation – a world rather as the Earth used to be in the Carboniferous Period, several hundreds of millions of years ago, when the Coal Forests were being laid down. Alas! not so. I can recall a lecture about Venus which I gave in London in the late 1950s; I made a series of twelve statements, every one of which was supported by the best available evidence, and every one of which turned out to be wrong. The surface temperature is of the order of 1000 degrees Fahrenheit, and all in all it is fair to say that the scene on Venus must be remarkably like the conventional idea of hell. (As an aside, I was once asked, during a television programme, to give my definition of hell. I gave it: bowling to a left-hander, on a dead wicket, with a Pakistani umpire. It wasn't quite what the questioner had in mind!)

I have said that it is only too easy to 'see' what one expects to see, and I once carried out a rather unkind experiment along these lines. At its most brilliant, Venus is in the crescent stage. Any small telescope will show its shape, and so will good binoculars, but can the phase be seen with the naked eye? There is evidence that it can. When Venus was suitably placed, I showed a photograph of it on *The Sky at Night*, and asked viewers to send in naked-eye drawings. With malice aforethought, the photograph I used was of the telescopic view, with the horns of the crescent pointing to the right. I waited to see what would happen.

Drawings came in thick and fast; in the end I had nearly five hundred. All except four showed the horns pointing right.

The remaining four viewers were decidedly baffled. 'I could see Venus as a crescent,' wrote one man, 'but it seemed to be the wrong way round. Are you sure you didn't put up the wrong picture?' Of course, these four were the only genuine sightings; the rest were due to imagination. And it does explain why so many people 'saw' canals on Mars once they had been reported.

I went back to Flagstaff, probably for the last time, a couple of years ago. Percival Lowell, Chick Capen and Clyde Tombaugh have left us, but they will not be forgotten, and Mars Hill will still be there in the centuries to come.

12 Xylophones and Transits

When was the first time that you handled a substantial sum of money? In my case it was in 1936, when I was thirteen. And thereby hangs a tale.

I had elected to spend sixpence a week on football pools, which were all the rage then (the Lottery lay far in the future). Of course, my mother had to fill in the form, and we agreed that if we ever won anything we would halve it. Suddenly, out of the blue, there arrived a cheque for £87, which was an absolute fortune. Her share went on paying bills. I said: 'Right. On my share we're all going for a fortnight's holiday in Belgium' so we went – my parents and myself. The train and sea fare from East Grinstead to Heyst-sur-Mer was £3, and the hotel in Heyst, ironically called the Hotel Cambridge, cost us five English shillings per day each – bed and breakfast and a three-course dinner each evening. My football pool covered the lot, including trips to places such as Ypres and Bruges, and when we arrived home there was still £7.10s left. I remember Mother saying: 'You've spent our winnings on us all, and what's left you must spend entirely on yourself.' That seemed fair enough, so I bought the little xylophone which I still have.

Music has always played a great part in my life, and I think I was tinkering at the piano before I could talk. I have never had a music lesson, and about musical theory my ignorance is complete, but I do have two things which are absolutely no credit to me:

perfect pitch and perfect time. This is something you can't learn, either you have got it, or you haven't. (On the debit side, I am hopelessly inartistic; if I see a superb picture in a majestic art gallery, my instinct is to look for the joke underneath.) Producing tunes never gave me any trouble at all, and neither did I care which key was needed; I could switch from one key to another instinctively.

I well remember one evening when I was eight. I was playing the piano, and a sudden thought came to me. 'This is silly. I can't read and write music!' So at the first opportunity, when I was in East Grinstead, I brought a sixpenny book, and taught myself the notation. Next I invested in a book of Viennese waltzes; Strauss, Waldteufel, Ziehrer and so on. Painstakingly I tapped them out from the music, and before long I was able to read reasonably well; I still have that book of waltzes. My main problem was that as soon as I knew the tune I could throw the book aside, and this is one of the reasons why I have always been a very poor sight-reader. The other reason is that because I am entirely self-taught, my keyboard fingering is clumsy.

My taste was, and remains, old-fashioned. I am no classicist; I am lukewarm about Beethoven, Bach and Mozart, though enthusiastic about Grieg and Tchaikovsky. Pop music I do not like at all. During my boyhood it was all jazz, with bands such as Henry Hall's. I preferred my waltzes, marches and polkas. When I began to compose I remained in the same period, and today my compositions belong to 1890 rather than 2003.

My mother, as I have said, was a singer who would have made her name but for the Kaiser's war. Music ran in the family. My uncle, Reginald White, was a barrister who abandoned the Law around 1895 and went on the stage, joining the D'Oyly Carte company and becoming a lead in Gilbert and Sullivan; Mrs. D'Oyly Carte was on record as saying that he was the best-ever Shadbolt in the *Yeomen*. He played along with the most famous professionals of the time such as Workman and Rutland

Barrington, and had a long career. (He was also one of the founders of the Cambridge Footlights.)

I was on the stage with him just once, long after he had retired and had gone to live at Richmond in Surrey. There were a good many ex-professionals in the area, and they put on shows which were up to any West End standard; once when they chose *Utopia Limited*, the last successful G and S, he played Scaphio, while I did Phantis. It was quite an experience!

My voice is all right for playing the Demon King in local pantomime, as I have done on countless occasions, but it is quite useless for anything else. (*En passant*, Lorna was tone-deaf, and equally lacking in a sense of time. On one of the rare occasions when we danced together, I inadvertently backed her into a cactus.)

That, incidentally, was one of the very happiest days of my life – within a day or two after my twentieth birthday. Most of the people for whom I cared most were in the same place – eight out of ten, more than ever before or since. The little band played Josef Strauss' waltz *Aquarellen*, and we ventured on to the dance floor. As a demonstration of dancing expertise it was not a success!

(A few weeks ago I was driving home, with the car radio on, when the BBC Concert Orchestra suddenly struck up *Aquarellen*. I have to admit that I was rather glad that for the next couple of minutes I was alone in the car...)

My first composition dates from the age of ten, and I claim that it was a perfectly respectable Viennese waltz. Many others followed. Then came the xylophone, and in fact my first foray on to the stage was when I was fourteen; I played a xylophone solo at a charity concert in the Whitehall Theatre, East Grinstead. I wasn't nervous, and I have to admit that when it comes to that sort of thing I have about as many nerves as the average rhinoceros.

Long afterwards, in the mid-1980s, I became the proud owner of a splendid British Premier xylophone. Ron Goodwin, the famous bandleader, invited me to play in a concert in Bristol, which was being televised. The audience was taken aback when I appeared, and even more taken aback when they realized that I could really play. In 1982 I was invited to play at the Royal Command Performance at Drury Lane, in front of the Queen; I played a solo of my own – *Hurricane* – and no eggs were thrown, so I assume it went down reasonably well; Her Majesty was amused!

People have been kind enough to say that I am not too bad a composer, and I have produced one tape and one CD of pieces I have written, but I am not a good performer myself. With the xylophone, drum and marimba I can hold my own, but not on the piano. My CD includes a Nocturne which is probably my best effort, but I have never managed to get through it without several false notes, so on the CD it is played for me by Robert Vallier, who is a concert pianist. I was honoured to be asked to write a new march for the band of the Royal Parachute Regiment; I called it 'Out of the Sky', and it is on their regular repertoire.

I did have one really weird experience. It was at a very sad time for me, in 1981, during my mother's final days. I had been asked to write something for the Liverpool Philharmonic, which was a new departure for me; it was to be a sort of a tone poem on the theme of Phaethon's Ride – in Greek mythology the story of the boy who drove the Sun-chariot for a day, and was about to set the world on fire before Zeus toppled him with a thunderbolt. Late at night I sat down at the piano, switched on a tape, and improvised. I remember nothing much about it, but twenty minutes later I 'came out of it' and wondered – what have I been playing? It was on the tape, and for the first and only time in my life I sounded like a good pianist. Hastily I re-taped it, because I knew at once that I could never play it again; it was too difficult for me. It took me about a month to write it down, note by note, and since then *Phæthon's Ride* has been played by symphony orchestras in various places (it is also on my CD, played by the Royal Scottish National Orchestra). Don't ask me to explain it; I can't.

On a rather different level, I recall a phone call I had just before

Christmas in 1998. The Youth Club was having its 'hop' with a pop band – and for some reason the drummer hadn't turned up; would I deputise? I did so, we went on until the early hours of the morning, and the teenage pop musicians admitted that I was quite capable of holding my own.

Musical instruments with strings, and those you blow, are not for me. I did once play a violin solo in public, but all the strings except one were made of rubber; the remaining string would produce a D natural – the only note I could actually play. I wrote a piece of music round it, and I was genuinely mistaken for a violinist. Asked to play again, I tactfully declined.

One of my xylophone experiences involved playing a duet with Evelyn Glennie. She is without doubt the best living percussionist, and how she manages it I know not, in view of the fact that she is totally deaf. We had played together once, and then she was caught on the BBC television programme *This Is Your Life*. I was the last guest; we had a xylophone ready, and we played the classic *Two Imps*. I wish I had one-hundredth of her skill.

Incidentally, she lip-reads so marvellously that you forget that she can't hear. She was once addressed as 'Evelyn', with a long E. 'No,' she said firmly. 'I'm Scottish, and my name is pronounced "Ever-lyn". How on earth did she know?

I was once conducted by Sir Edward Heath; I think it was 'Fossils', the xylophone solo in Saint-Saëns's *Carnival of the Animals*. He struck me as being a very good conductor, and I am told that in his youth he had to choose between music and politics; he may well have picked wrong! I have conducted only once, when the Oxford City Band was playing one of my marches and asked me to take over. It did not take me long to realize that I was out of my depth.

Do you remember Gerard Hoffnung? He was a fine musician, and also an unusual composer; I have played in his 'Concerto for Espresso Machine and Concrete Mixer' – on that occasion I was on the ocarina, which, in case you don't know, is made of clay and looks like a pear with a line of holes in it. The same concert included the 'Surprise Symphony'; there were plenty of surprises.

Avant-garde music defeats me. A few years ago I went to the penultimate night of the Proms, and one piece was written by an eminent modern composer Sir Harrison Birtwistle. In a subsequent radio programme I was asked what I thought of it, and I said, honestly, that to me it sounded exactly like a catfight. But neither am I a lover of grand opera; I vastly prefer Gilbert and Sullivan. Once, in Chichester, I took a part: King Gama in *Princess Ida*, the only lead I could play, because it needs a voice like a rusty saw ('O don't the days seem lank and long, when all goes right and nothing goes wrong?'). I have also composed three operas myself. The first, in my Holmewood days, was *Perseus*: I was responsible for the score and most of the lyrics (Jo Oldham contributed some, but I am not now sure which). In my version Perseus is far from the gallant hero of legend, instead he is an elderly old coward who once killed a Gorgon by mistake, and is nonplussed when confronted with a sea-monster sent by the god Neptune to ravage King Cepheus' realm. One character is Peleus, the Court Torturer and Protector of Public Morals; under Cepheus' liberal rule there were so few morals left in the whole kingdom, that the office became superflous. *Perseus* went down well, and much later I composed *Theseus*, again with a mythological background; in the legend Theseus kills the Minotaur, a horrible creature which ate people. Roger Prout, formerly of the Welsh National Opera and today very well known in the opera world wrote the libretto; our Minotaur was a delightful, kittenish little animal which liked biscuits. *Theseus* was put on successfully in Brighton and Chichester, and I hope it will go on again while I am still around.

My third and, I fear, my last opera was concerned with Galileo, the man who first turned a telescope toward the sky, and taught that the Earth is a planet moving round the Sun – for which heresy he was brought to Rome, condemned by the Inquisition, and threatened with torture and kept under house arrest until his death in 1642. In 1999 Roger produced Brecht's famous play, *The Life of Galileo*, and asked me to be the Narrator. It was then suggested that I might write a musical version, and the result was *Galileo: The True Story*. This time Galileo comes out very much on top. I explain that the real reason why the Pope became so angry was that Galileo's book is written in Latin – and the Pope has been trying desperately to conceal the fact that he cannot speak a word of the language. He has, however, definite policies:

> All trivial thoughts I will banish,
> And make sure that they never take shape;
> For it's only the hairs on a gooseberry
> That stops it from being a grape.

Certainly Galileo's accusers are in earnest:

> Of a crime of dreadful impact, Sir, you stand here now accused,
> Your scientific knowledge you have woefully abused.
> You'd wrench our noble Earth from its divinely ordered place;
> And send it whirling, madly through the blackness of deep space.
> You ridicule Holy Writ and make it but a mock,
> And replace our Creator with the dreadful Mr. Spock.
> What say you, foolish miscreant, before the torture dire
> Descends upon your hapless frame in to streams of liquid fire?

You can see the type of verse that I write! If you want to try out the Grand Inquisitor's song, here it is …

I am the Grand Inquisitor,
My power is absolute;
I keep an eye upon all men,
And women too, to boot.

To challenge me is far from wise,
Because I don't play fair;
And this is really no surprise –
To you, I say Beware!

I have my thumbscrews all prepared,
It really is great fun;
And when the victim's nicely trussed
I'll know the game's begun.

I'm not a very pleasant cove,
I'm sure you will agree;
My grin will make your blood run cold
To my sadistic glee.

I hold the power in my hands,
My ways are all my own;
And it is music to my ears
To hear my victims groan.

When all the fun's about to start
You bet I'll be right there;
The luckless wretch who crosses me
Will soon know deep despair.

You may well think that I'm a swine,
And you will be quite right;
I have no sense of decency –
I am jet black, not white.

So I, as Grand Inquisitor,
Will keep you on your toes;
I have no mercy – not a scrap,
For either friends or foes.

Galileo was produced at the Arts Theatre in Cambridge, in March 2002, by enthusiastic students, mainly astronomers; the moving spirit was Chris Lintott, in the middle of his degree, who had joined me in *The Sky at Night* and with whom I had co-authored a book (*Astronomy for GCSE*). It ran for a week, was packed out every night. During the last week in August 2002, the Cambridge cast brought *Galileo* to Chichester, and again it was a great success. Finally, on 1 September, they gave an evening performance in my garden – to an audience of about 100! For the previous five days the entire cast had been staying with me (with rooms and tents, we slept twenty-seven) and it was sad when the last show was over; the cast dispersed, and went their different ways. Farthings seemed eerily quiet when they had all gone.

Two musical links with *The Sky at Night* come to mind. The first dates back to 1960, when we decided to tackle transits of Venus. (I forget why we chose that particular moment; anyway, we did.) Venus, remember, is closer to the sun than we are – on average 67,000,000 miles out, as against our 93,000,000 – and there are times when Sun, Venus and Earth line up, with Venus in the mid position. Venus then appears as a black disk in transit across the brilliant solar disk, large enough to be seen with the naked eye – so I am told; I have never seen a transit, and neither has anyone else now living (2003), because the last one occurred in 1882, and anyone remembering that would have to be decidedly advanced in years. Transits occur in pairs, separated by eight years, after which there are no more for over a century. Thus there were transits in 1631, 1639, 1761, 1769, 1874 and 1882; the next are due in 2004 and 2012.

For our programme, there were two suitable venues; Hoole in Cheshire (where Jerimiah Horrocks made the first recoded observation of a transit, in 1639), or Tahiti in the South Seas (where Captain Cook observed the 1769 transit, during the voyage which led on to the discovery of Australia). Hoole or Tahiti? Being the BBC, you can guess which site was chosen, and we set off for Hoole.

Jeremiah Horrocks was a brilliant young mathematician, who alone had predicted the 1639 transit (that of 1631 had passed unnoticed, as it occurred during the night over Europe). Horrocks set up his telescope ready for the event, due on 4 December. In fact he did see it, and made some useful measurements. The only other successful observer was his friend William Crabtree, from Manchester. (Horrocks' brother Jonas, at Liverpool, was also on the alert, but was clouded out). Go to Manchester Town Hall, and you will see a large mural showing Crabtree observing the transit. I am afraid it is not very accurate, but at least it is a nice picture.

It has been said that Horrocks was a curate at Hoole. This seems questionable, but at any rate there is a stained glass window in Hoole Church commemorating the transit, and there is no doubt that Horrocks observed from the garden of Carr House, in the village. When we went there we found that Carr House had been turned into a doll museum; I found it slightly sinister, but we were made welcome, and we set up for transmission – live, of course; recorded programmes lay in the future.

Half an hour before we were due on the air, we made a shocking discovery. The record of 'At the Castle Gate', an old 78 rpm vinyl, had been left in London, and there was no time to fetch it. Instead, we looked round for a piano. There wasn't one, but we unearthed an ancient harpsichord. I had never played a harpsichord, and I had never played 'At the Castle Gate', but clearly I had to do my best, and I bravely played us in and out of the programme. I hope I didn't hit too many false notes; it wasn't recorded, so I will never know.

On one other occasion I entered the musical lion's den. This was in 1981, a hundred years after William Herschel discovered the planet Uranus. We were presenting a programme about Herschel, and we were doing it from No. 19 New King Street in Bath, where Herschel was living when he made his great discovery. Up to that time he had been a professional musician, and was organist at the Octagon Room; he was a composer, not of the first rank, but perfectly acceptable. I borrowed a manuscript from the Royal College of Music, and played it during transmission. It was not difficult; people have commented that Herschel as a composer is a little like Mozart gone stale.

I think I will digress to say more about No. 19 New King Street, because I was deeply involved in its subsequent history. It is not in the best part of Bath, but it is easy to find (assuming that you can fathom the intricacies of the one-way traffic system, which is one of the most confusing in the British Isles; its only possible rival in this respect is Worcester). In 1976 I had a phone call from Miss Philippa Savery, a resident who was not an astronomer in any sense of the term, but was keenly interested in all affairs pertaining to the city. Did I know about No. 19 New King Street, and did I know that it was threatened with demolition?

I did know about No. 19, which was the only remaining Herschel house, his final home, in Slough, had been pulled down in 1961 – if it had not been it would have fallen down anyway because it was riddled with wet rot, dry rot, woodworm and practically everything else. (I went there on the day before demolition started, and took pictures.) So only No. 19 remained.

I drove to Bath, met Philippa – who became a great friend – and went to the house. Obviously we were batting on a very difficult wicket. The house was in bad repair, and the garden was a wilderness. We decided to take a mad gamble: we hired the Octagon Room, called a public meeting and launched the Herschel Society. We contacted the local radio station and the local paper, we put up notices, and waited. Candidly, I didn't think we had the ghost of a chance.

To our delight, that first meeting was well attended, one of the speakers was Colin Ronan, the well-known historian of science, who was a close personal friend. Everyone was enthusiastic and we raised enough cash to keep the demolition squad at bay. Then a local resident, Dr. Hilliard, stepped in and bought the house on behalf of the City – truly a magnificent gesture. One thing led to another; further support was forthcoming, and today No. 19 has become a small but excellent Herschel Museum. I can assure you that it is worth visiting next time you find yourself in Bath – always provided that you can defeat the one-way traffic system!

13 The Return of the Comet

During the 1970s and 1980s I was a frequent flyer across the Atlantic, because I was generally present at NASA's Mission Control during the planetary missions. Mercury, Jupiter, Saturn, Uranus and Neptune were all by-passed and each encounter produced its quota of surprises. One regular companion on television was Dr. Garry Hunt, who became (and remains) a close friend – one of NASA's Principal Scientific Investigators, and as a television presenter second to none. I remember one of our conversations in 1980, when we were at NASA and the Voyager 1 space-craft was approaching Titan, the largest satellite of Saturn, which is larger than our Moon and actually larger than the planet Mercury, though less massive. Titan has a dense atmosphere, but was it dense enough to hide the surface features? I doubted it; Garry maintained that all we would see were cloud-tops. He was right, and I was wrong: the pictures showed us nothing more informative than orange smog. We are still not very clear about what Titan is really like, but we may find out in 2004, when a new probe, Cassini-Huygens, arrives there. It has been on its way ever since 1996, so that it has taken its time; it has to follow a roundabout route, rather like a train going from Brighton to Bognor Regis by way of Penzance.

Generally speaking, our television programmes went well. Alas, this was not true of a programme about the return of Halley's Comet, in 1986, which proved to be a total disaster.

Comets are the most erratic members of the Solar System. They are quite unlike planets, and indeed a comet has been described as a dirty snowball which releases gas when nearing the Sun, so that it develops a coma as well as one or more tails. Many small comets move round the Sun in periods of a few years, so that we always know when and where to expect them, but really bright comets move differently; they come from the far reaches of the Solar System, and their paths are very eccentric, so that their orbital periods may be centuries or even many thousands of years. Therefore we cannot predict them, and neither can we predict how they are going to behave. In 1973 one comet, discovered by Dr. Lubos Kohoutek and named after him, was expected to become magnificent, but in the end it was very feeble indeed, and was widely termed 'the flop of the century'. It may be better when it pays us its next visit, in about 75,000 years' time.

The only bright comet which comes back regularly is Halley's named after Edmond Halley, the second Astronomer Royal, who was the first to realize that it is predictable. It has a period of 76 years, it was seen in 1751, 1835 and 1910, so that it was due once more in 1986. Preparations were made. In London, Brian Harpur founded the Halley's Comet Society, which had no rules, no aims and no ambitions; all it did was to meet sporadically, always on licensed premises – I recall one notable gathering in the Long Room at Lord's. When the comet returned we did organise a charity concert, during which almost everything went wrong (I was lucky in as much as my xylophone solo was the opening item, before the lights failed for the first time). We did raise a good sum, despite a demonstration outside made up of people who were firmly convinced that the comet presaged the end of the world. Otherwise, I have to admit that the Halley's Comet Society was the only completely useless organisation in the world, apart of course from the European Parliament.

The 1986 return was eagerly awaited, because for the first time it was possible to send spacecraft to rendezvous with the comet. The Americans pulled out on the grounds of expense, but the Russians sent two missions, the Japanese two, and the European Space Agency one – called Giotto in honour of the Italian painter Giotto di Bondone, who had shown the comet in his famous picture 'The Adoration of the Magi'. He was wrong in claiming that the comet was identical with the Star of Bethlehem, but the name seemed attractive; Giotto was scheduled to plunge right into the head of the comet, and send back close-range images of the icy nucleus. The Russian and Japanese spacecraft were fly-bys.

The BBC made great preparations for a really epic programme. The headquarters of the European Space Agency were at Darmstadt in what was then West Germany, and it was here that the leading experts gathered, including Jan Oort of Holland and Fred Whipple from the United States, who really knew more about comets than anybody else; also present was Roald Sagdeev, from the USSR. It seemed sensible to do the programme from there, but unfortunately *The Sky at Night* has a very low budget, and the prestigious *Horizon* programme came in, with the obvious intention of taking over. I was able to go to Darmstadt with the *Sky at Night* team, but the main programme was to be controlled from London.

At that time there was still some uncertainty about what a comet's nucleus was like. Fred Whipple's dirty snowball theory was favoured, but there was still some support for a theory due to R. A. Lyttleton, of Cambridge, who regarded the comet as a 'flying gravel-bank', made up of particles which jostled together when nearing the Sun. It was hoped that Giotto would clear the problem up. I had seen the launch, from Kourou in French Guyana and on 6 March 1986 I arrived at Darmstadt, ready for the programme a week later. I was particularly interested because in my rôle as an astronomer I was also a member of the International Halley Watch, and I had been photographing and studying the comet as intensively as I could.

Darmstadt was a hive of activity. There were arguments, too, because Giotto carried a great many different experiments, and the various teams had different requirements. Some of the experimenters wanted to go as close to the nucleus as possible, while others preferred to stay further out. In particular the camera team, led by Dr. H. Keller of West Germany, wanted to go no closer than 600 miles. Eventually a compromise was reached: just over 300 miles. The actual minimum range was 375 miles, and it transpired that Giotto arrived only four seconds late, after a journey which had started in the previous July.

It was clear that the main danger to Giotto would come from the impacts of dust-particles. Halley's Comet orbits the Sun in a retrograde or wrong-way direction, so that any debris from it would meet the probe head-on at very high speed – and under these conditions even a tiny dust-particle will have tremendous destructive power. Bumper shields, designed by Fred Whipple, had been fitted, but nobody knew how effective they would be. I spoke to Fred a few hours before the encounter, and he remained confident that Giotto would survive, but other investigators were less sanguine. Most people expected to see an icy nucleus with spurting jets and streamers; there was little support for the views of Sir Fred Hoyle and Professor Chandra Wickramasinghe, who claimed that the comet would have a very dark nucleus coated with organic materials.

Just a few minutes from the start of transmission I had a phone call from Chandra, in Cardiff. Would I stress that he and Fred had forecast a dark nucleus rather than a bright one? I promised to do so, and I kept my word, though at an early stage the whole programme became decidedly chaotic.

The first essential was to find out just where the nucleus was;

it was completely hidden by the coma, and could not be seen at all. This was where the Russian and Japanese spacecraft helped. They by-passed the comet well before Giotto made its sortie, and their results were of the utmost value.

We began the programme – live, of course – on the late evening of 13 March, and things soon started to go awry.

At Darmstadt I could call on the world's leading experts, including Fred Whipple, Jan Oort and Roald Sagdeev. However, at Greenwich (why Greenwich) *Horizon* had assembled a team of astronomers, none of whom had ever shown any particular interest in comets.

Nobody could tell what would happen when Giotto plunged into the comets head and was bound to be battered by debris. In fact the first dust impacts were recorded at a distance of 175,000 miles from the nucleus, but not until the range had been reduced to 5,000 miles was there an impact violent enough to cause any alarm. Fourteen seconds before closest approach, when Giotto was still 1,058 miles from the nucleus, there was a direct strike from a piece of material about the size of a grain of rice. The spacecraft was jolted out of alignment, and the signals became intermittent. I began to explain, but I was unable to do much in the brief intervals when Greenwich allowed me 'on the air'. We had expected the main results to come from the Halley Multi-Colour Cameras, and we were not disappointed, but there were some major surprises. First, the nucleus was much warmer than had been predicted. In shape it was likened variously to a baked bean, an avocado or even a banana; it was about 9 miles long by 5 miles wide, with a mass of from 50,000 million to 100,000 million tons. (This may sound a great deal, but it would take at least 60,000 million comets of this mass to equal the mass of the Earth.) The low density indicated that the comet was made up of 'fluffy' material, with water ice as the main constituent. There was considerable surface structure, including a bright region which seems to have been a hill almost a mile high; there were craters, jets, and three main active areas. All the jets appeared to issue from one small area on the sunlit side of the nucleus, and the remaining 85 per cent of the surface was comparatively inert. The rotation period was given as 7.3 hours, but the nucleus was behaving rather in the manner of a spinning gyroscope. It was estimated that the comet would have lost about 300,000,000 tons of material before moving back into the remoter parts of the Solar System and becoming inactive once more.

The nucleus was dark, just as Fred and Chandra had predicted; Dr. Keller compared it with 'black velvet', reflecting no more than 2 to 4 per cent of the sunlight falling upon it. There were two bright patches, each about three-quarters of a mile across, which had been recorded by the Russian fly-by probes and had given the misleading impression of a double nucleus. As we now know, this was due to the presence of an upper layer of dark dust, which prevented the ice below from evaporating quickly. Fred Whipple's 'dirty snowball' contained more dirt than had been anticipated.

When we came off the air, at the end of transmission time, I was not at all pleased. However, there was more to come, even if not on television. The comet was still there, and I wanted to take full advantage of it, as I was unlikely to be around at the time of the comet's next visit, in 2061. The best viewing times were expected to be in March and April 1986, from the southern hemisphere, and I joined the party which the Explorers Travel Club was sending to Australia. I planned to photograph the comet as often as I could, and use the pictures in a later *Sky at Night* programme when we had had time to pull all the results together.

Our party was about forty strong. We flew to Sydney, and from there I made a couple of broadcasts; I then drove to Alice Springs, in the middle of the Australian continent, which is largely Aborigine. I arrived there on 11 April. There had been no appreciable rainfall for five years, and none at all during the previous

With Mother, at Farthings. Believe it or not, she was then 87.

Selsey Cricket Club 1st XI, 1973.

Farthings, 2000.

Dish of the Lovell Telescope at Jodrell Bank; figures painted on it for the 30th anniversary.

Me with Bonnie, at Selsey. She lived to be 19½.

Xylophone practice before a concert – 1997.

Chess tournament, Hastings 1996. I forget who won that game.

My Party coat – when I bought it in Kuala Lumpur I thought it was Oxford blue . . .

Another musical moment. . . practice makes perfect.

Selsey Pantomime: the Demon King.

In my Farthings study. Note the Woodstock on my desk.

Practice leg-spin in the garden before a match.

Invitation to my Sky At Night 40th anniversary party - by Paul Doherty.

eight months – yet the instance I set foot in the town, the heavens opened. The local Pressmen were quick to blame me, and solemnly photographed me as I shielded myself under an umbrella which someone had rooted out specially for my benefit.

The rain soon cleared, and the comet shone forth. By now it was moving against the stars of the Centaur, which is too far south to be seen from England, but is quite unmistakable. Together with Douglas Arnold, who is (unlike me) an expert photographer, I drove out into the desert so as to get as far away from the town lights as possible. We were not alone, because several of the local astronomers accompanied us and acted as our guides, but the trip was not uneventful. When we were unwise enough to leave the main road and take to an unmade track, we became acutely aware that the wheels of our car were sinking into the deep red dust, and it took several hours to extricate them.

Following an inspection of Ayers Rocks, a huge block which rises from the landscape for no apparent reason, we set off in the direction of Darwin, camping at various sites. It was on 17 April, near Tennant's Creek, that I had an odd experience. By about four o'clock in the morning the comet had dropped so low in the sky that no further photography was possible, and I decided to get some sleep, so I went into my tent and undressed. Suddenly the flap opened, and in came – a wallaby which wanted a drink! I gave it two cupfuls of water, which it swallowed happily and then withdrew. (This is not such a tall story as it may sound. We were on an official camping site, and the wallaby was used to human company, so that one could go up to it at any time and tickle its ears.)

By then the comet was in Centaurus, and people kept on coming up to me to say that there were two comets, side by side. The second object was, in fact, not a comet at all, but a globular cluster, Omega Centauri, made up of about a million stars and so far away that its light, moving at 186,000 miles per second, takes 17,000 years to reach us. Still, the two did look very much alike, and for a while they were in the same binocular field.

I came back by way of Bali, in Indonesia. On 24 April there was a total eclipse of the Moon, and for more than an hour the sky darkened, so that the comet could be seen with the naked eye for the very last time. It was an unforgettable sight, particularly in that romantic setting. In a way I felt sad, but it was an ideal way to say farewell to a fascinating visitor whom I will never see again.

14 Chasing Shadows

Eclipse-chasing can become addictive. As I have said, there is nothing in Nature to match the glory of a total eclipse of the Sun, and it is something never to be missed. My first eclipse was that of 1954, in Sweden, followed by the hilarious episode of 1961 and the mountain oxen. For the next few years I was unable to travel far – this was before and during my spell at Armagh – but once I became a freelance again I could look around, and I set my sights on Siberia.

Most people think of Siberia as a land of fur-capped peasants, salt-mines, tundra and reindeer, seared by Arctic winds and permanently at sub-zero temperatures. This may be true of the Asiatic part of it – not having been there, I don't know – but the European side of it is less forbidding, and it was from here, on 22 September 1968, that another total eclipse was due. Flushed with our earlier *Sky at Night* success we laid our plans, only to find that taking television equipment across the Siberian border was impossible. Not that the Russians were unhelpful; they weren't, but the purely practical obstacles defeated us. Therefore, it was agreed that I would go on my own, take what pictures I could, and give a first-hand report when I came back.

As a piece of personal research, I proposed to carry out a hunt for comets. These bizarre members of the Sun's family can often creep up on us unawares, and if they approach from behind the Sun we can easily overlook them. When the eclipse becomes total the sky darkens enough for any errant comets to show up, as had happened in 1882 when a photograph showed not only the solar corona but also a bright comet, a few degrees from the Sun's edge, which had never been seen before and was never seen again. Of course it was a 'long shot', but it was worth a try.

Yurgamysh, the site chosen by the Russians for all the various expeditions, is not far inside the Siberian border, and quite a number of enthusiasts, both amateur and professional were scheduled to converge there. The nearest towns you are likely to find on an ordinary map are Kurgan and Sverdlovsk. From Keele University came an old friend, Dr. Ron Maddison; from Holland, Professor Houtgast; there were Swedes, Danes, Americans, Germans – in fact representatives from many nations. I found out that the weather was unlikely to be either wildly hot or bitterly cold, because that part of Siberia was going through its autumn, which lasts for about a week.

Then came the first hitch. The Soviet authorities became very difficult about granting me a visa. Russia itself was fine – after all, I was an Hon. Member of the Astronomical Society of the USSR, thanks to my Moon-mapping activities – but Siberia came under a different set of rules and regulations, and, believe me, our own charming Civil Service had nothing on that of Soviet Russia. Sir Humphrey Appleby would have been in his element there (I doubt whether it is very much better today).

Once this became known, the Press questioned me closely, but I said little except that the Russians were fully entitled to say who they did or did not want in their country. If I were regarded as a filthy Western warmonger, who had made many ideological mistakes and had deviated strongly from the Socialist doctrines of Karl Marx and V. I. Lenin, who was I to argue? I cast my mind back to my first visit to Moscow, in 1959, when I almost caused an international incident by parking my fur hatski on a bust which turned out to be that of Lenin, but I hardly thought that that could be the cause of the problem. All I could do was to wait and see.

Four days before the eclipse I had a call from the Royal Society. Urgent messages had come through from the USSR Academy of Sciences in Moscow. I was welcome in Siberia after all. Could I catch the flight from Heathrow at noon on the next day, having first been to the Soviet Embassy to collect the visa waiting for me?

It was all rather a rush. My car was being serviced; there were no reliable trains to London, because as usual British Rail had been caught completely off its guard by the onset of autumn, and the roads were little better, because many of them were flooded. Somehow or other I got to London by breakfast-time, picked up my visa, prevailed on the BBC to provide some travellers' cheques, and struggled to the airport. By noon I was on an Aeroflot flight to Moscow.

End of the problem? Not so. My visa turned out to be for Moscow only. Luckily I managed to contact Dr. Khurikov, of the Soviet Academy, as soon as I landed and he explained that a Siberian visa would have to be rushed through. Normally, to rush anything through the Russian bureaucracy takes about six months. Dr. Khurikov, to his eternal credit, managed it in about six hours, and again entrusting myself to Aeroflot I travelled onward to Kurgan, where I joined up with some Danish scientists before setting off for the site at Yurgamysh. There was one odd incident at the Kurgan hotel. I wanted to contact Ron Maddison, who was presumably already at this eclipse site; the Russians misunderstood me – was I ill, and if so should they call a surgeon? Before I knew what was happening, the hospital authorities were on my trail. I had asked for Dr. Maddison; they thought I wanted a doctor of *medicine* ...

Having sorted that one out, we went on a 2½-hour drive to Yurgamysh, which consisted entirely of a children's summer camp which had been evacuated a few days earlier because of the impending onset of winter. The expeditions were setting up equipment in a large field, and as I approached I could see plenty of activity. In the hurry of departure I had had no time to pack anything except a single suitcase, which I was carrying and which was crammed with cameras to the exclusion of almost everything else; I did not even have an overcoat, and as I walked across to the site asking 'Have I come to the right place?' there was a good deal of ribald laughter.

The equipment was quite elaborate. The Russians meant to make studies of the sky brightness and the structure of the corona. The Americans were concentrating on the photometric measurements of the chromosphere; the Italians, the Swedes and the Swiss were particularly interested in a phenomenon known as the flash spectrum. Ron Maddison's goal was large-scale photography. Then, as I went around, I came across the famous Dutch astronomer Jan Houtgast, whose equipment seemed to consist solely of a large armchair. Knowing that he was one of the world's leading eclipse experts, I could not resist stopping to ask him what he meant to do.

He gave me a wise smile. 'I have been to many eclipses,' he said in his fluent but accented English. 'I have made many observations. Never have I had time to watch an eclipse and enjoy it myself. So this time – I do absolutely nozzings!'

Neither did he. While everyone else was working flat out during the fleeting moments of totality, he simply laid back in his armchair and watched. I am sure that he had the right attitude of mind.

Conditions proved to be quite good. There was no cloud trouble; the corona was spectacular, and so were the prominences. I swept for comets, but I have to admit that for part of the time I followed Professor Houtgast's example and merely stared. We had expected 43 seconds of totality, but actually we had only 37, because the Sun began to reappear early through a valley in the Moon's limb. The light flooded back over the Siberian field, and the sensation of unreality – so evident before totality – had passed.

I think we all felt elated. The experiments had worked well, and it had been worth coming so far across the world.

I do have one other vivid memory of that trip. The Russians had provided a banquet in the restaurant of the children's camp, and on the evening after the eclipse invited all the astronomers, plus the entire permanent population of Yurgamysh (about ten people all told). Vodka flowed freely, and there were speeches, including one from Dr. Gnevyshev, leader of the Soviet team, and then brief statements from the various national parties. When my turn came, I stood up and said how glad we all were to be here. 'It's really great,' I said. 'All of us are here in friendship, concerned entirely with the Sun. If only the politicians could see us now!' At which a leading Soviet astronomer, Dr. Vshekesviatsky, rose to his feet, brandished a glass of vodka, and thundered: 'Ah! The politicians! We do not want these gentlemen here!'

He didn't actually say 'gentlemen' – the word he used was much more basic – but he had the right idea. I couldn't have put it better myself.

The eclipse of 30 June 1973 was widely called 'the eclipse of the century', because the length of totality was over seven minutes, which is almost the maximum possible. For *The Sky at Night* we did not want to miss it, but it meant going abroad, because from England only a tiny segment of the Sun would be obscured, and casual observers would be unlikely to notice anything at all. The track of totality began off the African coast and then crossed the Dark Continent, passing through the states of Mauritania, Mali, Niger, Chad, Sudan and Kenya before tapering off in the Indian Ocean. At that stage I knew very little about any of these countries, but we had to make a decision, and we investigated as carefully as we could. Patricia Owtram had taken over as producer of the programme, and there could not have been a better choice. Women science producers came in two distinct classes, the very good and (more often) the very bad; Pat Owtram came into the former category!

From the point of view of climate and length of totality, Mali seemed to be ideal. (Do you know just where Mali is? I didn't, until I looked it up.) Unfortunately, there were some marked drawbacks. Roads and communications in general were said to be conspicuous only by their absence, and there was also the violently left-wing Government to be considered. We were warned that although we ought to be able to get into the country, it might be much more difficult to get out again, and there was a danger of our being impounded as dangerous capitalist agents. Prudently, we crossed Mali off our list.

The Cape Verde Islands, off the African mainland, were much more inviting, but after consulting the weather-men we established that complete cloud cover at that time of the year was practically certain, and, reluctantly we abandoned Cape Verde. We next looked at Kenya, but the length of totality was shorter, and the meteorological outlook was rather uncertain. (In the event, a major expedition did go there, only to be met by a hostile reception from the local inhabitants, who were convinced that the white men had arrived in order to put out the Sun.) Next on the list was Mauritania, but again facilities were poor. Finally, Pat Owtram came up with the idea of transmitting directly from the deck of a ship.

Here too there were obvious drawbacks, but there was a good deal to be said on the plus side, particularly as the good ship *Monte Umbe* was going to the centre of the track in any case, carrying a couple of hundred mad astronomers; the expedition had been organized by the Explorers Travel Club, and I had been invited to join it. So we settled for the *Monte Umbe*, knowing full well that rough seas at the vital moment would prove a real hazard.

But how could we get our films back to the studio in time for transmission the same evening? We could arrange for a linkman in the Television Centre studio, but the films and my on-the-spot commentary were essential. The only answer seemed to be to get the material flown back direct from the nearest port, which was, of course, in Mauritania.

Mauritania itself is almost entirely Sahara Desert, and has only two towns of any size. One is the capital, Nouakchott, which then consisted entirely of tents (I gather that the Prime Minister had a bigger tent than anyone else). The other is Nouadhibou, at the very edge of the coastal desert, and I asked whether the BBC had any representative there. The date, please remember, was 1973.

'Yes,' was the reply. 'We do have our man in Nouadhibou, and no doubt he'll be able to help.'

'Can we telephone him?' I asked.

This seemed reasonable enough, so we tried; in fact we tried for a complete morning, with a total lack of success. All we got down the line was a series of depressing squeaks. Finally, I became somewhat suspicious. 'When did the BBC last contact our man in Nouadhibou?'

The answer was... 1948!

We consigned him to outer darkness and managed to get in touch with the local officials, who did their best. We also called in a friendly pilot, and in the end it all worked out well.

A week before the date of the eclipse we trooped on board the *Monte Umbe*, joining many old friends – mainly amateur astronomers, but a sprinkling of professionals as well. We were next introduced to the captain, who was Spanish and who was unfailingly helpful and courteous. His name was Captain Mirallave, though I am quite sure that he was really the Duke of Plaza-Toro in disguise. When we asked how close he planned to go to the centre of the eclipse track, he gave a knowing smile. 'The waters in this part of the ocean have not been well charted, he said. 'There are two maps, one French and one Spanish. The French map shows many dangerous rocks. The Spanish map shows no rocks. I will be using the Spanish map.'

Well – how could I, as a mere airman, make any comment? It was, however, reassuring to know that one of the amateur astronomers on board was Commander Henry Hatfield, R. N., who had been the Navy's leading hydrographer. In the end the *Monte Umbe* did get to exactly the right place at exactly the right time, avoiding all contact with the French rocks. Full marks to the Duke of Plaza-Toro.

We worked out our programme, in which we featured the other eclipse parties. One expedition, of special interest, involved a Concorde aircraft. The Moon's shadow flashes along at a tremendous rate, which is why the duration of totality at any part of the Earth's surface is so brief, but Concorde had enough speed to fly under the shadow, keeping pace with it. The plan was to take off from Las Palmas, in the Canary Islands, and fly along at a height of 55,000 feet. This meant that most of the dense layers in the atmosphere would lie below, and the Concorde observers could expect to have a 'totality' lasting for over half an hour; enabling them to make continuous records and note any short-term changes in the corona and other features – something which had never been done before (though it is easy enough today, with 21-century type equipment). Dr. John Beckman, one of our regular *Sky at Night* guests, was on the Concorde, and promised to let us have his results as soon as he knew them himself.

The Concorde had to be modified to some extent, and three holes were bored in the roof to accommodate the astronomical equipment. One thought occurred to me. I knew that a team based inland from Nouadhibou was planning to launch an upper-atmosphere rocket during totality, and I had a fleeting fear that if they dispatched it at the wrong moment there might be four holes in the Concorde. Luckily, these misgivings turned out to be groundless.

The journey to the site was most enjoyable, though I did go through one minor crisis. We stopped off for a few hours at Las Palmas, and were able to look inside the waiting Concorde and take some film of it; I then went into the town (which qualifies as a concrete jungle these days) to telephone my mother, on her eighty-seventh birthday. I hired a taxi, and set off. I made my call – in fact I got straight through; much easier than ringing up Kent from Sussex – and re-entered the taxi to go back to the dock. The driver accelerated into the main road, and crashed into another taxi. Disaster! Police swarmed round, pointing their fingers at me. 'Witness, witness,' they seemed to be saying, and I had to do some quick thinking. If I were held up in Las Palmas I would miss the eclipse, because I could hardly expect either it or the *Monte Umbe* to wait for me.

I cannot speak Spanish, so I stood back. While the police were arguing with the taxi-drivers and with each other I dashed round the corner of the street, hailed another cab, and demanded to be taken straight to the dock. Mercifully, the driver obeyed. I lost no time in getting back on board, and stayed prudently in my cabin until we sailed a couple of hours later. Fond as I am of the Canaries, on this occasion I was glad to see them receding in the distance.

By 30 June we had reached a point some twenty-four miles from the Mauritanian coast, right on the central track. The main problem, of course, was stability. A swaying deck is not the ideal place to set up a telescope or a camera, and the various participants had all sorts of devices to counteract ship movement. Some of the instruments were mounted on gimbals; one telescope was fixed to a piece of mechanism which looked like a demented windmill; weights, pendulums and other ingenious dodges were much in evidence. I had considerable respect for Horace Dall, the famous optical expert, who scornfully rejected all such things and merely planned to lie flat on his back, balancing his camera on his nose. I had no telescope myself, because I knew from the outset that I would be fully occupied with my television commentary.

Elaborate rehearsals were held, to the bewilderment of the ship's crew. I well remember the efforts of the hostesses and organizers to marshal us. 'There will be deck games at 10.30 this morning....' And shortly afterwards, 'There will be no deck games this morning.' Bingo and other attractions fared no better. I felt particularly sorry for the ship's photographer, who found to his despair that he was unwanted, because everyone on board had at least three cameras. Even on the morning of the eclipse deck games were still announced, but it would have been rather difficult to hold them, because every square inch of deck space was occupied.

I also remember the Case of the Orbiting Professor. We had set up a rehearsal sequence, and were about to confer with Commander Hatfield when we saw a large figure zooming in toward us – that of an American professor who believed in taking regular exercise, and made a habit of achieving twenty orbits round the main deck before breakfast (you could even set your watch by him). He was just about to blunder into our line of sight when Patricia Wood spotted him, and made a wild lunge at him. One of our sound assistants reached out to grab him, but too late; he continued on his way, and then, realizing the enormity of his crime, stood right in front of the camera and took ten minutes to apologize. As our rehearsal time was strictly limited, we all felt an uncontrollable desire to tip him overboard.

On eclipse day itself the weather was clear, and, mercifully, the sea was so calm that the various stabilising devices were not needed. There was a feeling of tremendous tension as the light faded; then came the onrush of the Moon's shadow, and the sky turned a curious mauve colour. Two planets could be seen, Venus below the Sun and Saturn above, though I could not make out any stars.

With over seven minutes totality, we had a real chance to draw breath, and it was all very different from our fleeting 37 seconds at Yurgamysh. I concentrated on the commentary; beside me the camera team, led by Philip Bonham-Carter, was hard at work getting the best possible shots. Then, suddenly the diamond ring appeared, and the sunlight came back as the shadow of the Moon sped away across the ocean.

It was then that I realized that I had dropped my small hand-camera. I stooped down to pick it up – and my trousers split across the back. It was my bad luck that one of our photographic experts happened to be ideally placed to 'shoot' me. He later showed me the picture, I never managed to obtain a copy of it, but at least I must be one of the few people on record as having split his trousers during a solar eclipse. (Yes, Claude; I said *split*.)

There were a few mishaps. Someone had a faulty film, and there were a couple of cases of camera jam. One luckless enthusiast took forty photographs of the corona – with his camera hood in place all the time; when he found out, he was seen walking purposefully toward the side of the ship, though somebody pulled him back before he reached the edge. However, in general the results were good, particularly Horace Dall's. Not so with the Nouadhibou rocket, which had been sent up to take pictures of the corona in ultra-violet light; the de-spinning mechanism of the rocket refused to work, and the whole film was blank. A team on Mauritania had mixed fortunes, partly because of a sandstorm and partly because the nice, friendly inhabitants surrounded the members of the expedition just before totality and proceeded to hurl rocks at them. In Niger, a violent duststorm blacked out the eclipse completely. On the credit side, the Concorde flight was faultless, and John Beckman and his colleagues achieved everything that they had set out to do.

With totality over, our first need was to get our films back to London. We went into Nouadhibou, contacted the pilot, and waved him off. Then we had a few hours to look around the town itself, which I can honestly say is the most ghastly place I have ever seen. The only industry takes the form of an evil-smelling fertiliser factory; there are flies everywhere, and the sand gets into your mouth and up your nose. Add the scorching heat, the glare and the universal dirt, and you will appreciate what I mean.

I recalled all this during the ship's concert, given on the last evening of our voyage home and just before we bade a regretful farewell to the *Monte Umbe*. Normally, the final concert is presented by the entertainment staff, but this time we took it over. My own contribution, right at the end, consisted of a music-hall turn ending with a song about Nouadhibou, which I wrote and composed specially for the occasion and which was given its one and only public performance. I reproduce it here just in case anyone wants to try it out at a concert in the Royal Festival Hall.

We walked into the desert sun,
The day had only just begun,
We'd heard such glowing rumours of the place.
We looked for sheikhs on Arab steeds,
And women dressed in gorgeous beads,
But when we saw what we had got to face –

(Chorus)

Boo, boo, Nouadhibou,
We're glad you're far away.
True, true, Nouadhibou.
You told us we could stay.
But pooh, pooh, Nouadhibou.
It's time to say 'Good-day';
So toodle-oo, Nouadhibou.
Shall we come back? Nay, nay!

We watched with glazed and goggling eyes
As multitudes of desert flies
Surrounded us and nibbled at our coats.
We dodged the crowd of market boys
Who tried to sell us junk and toys,
And cursed us as we ran back to our boats:

(Chorus)

And as we leave this sandy land
It isn't hard to understand
That we will come here never, never more.
One visit here was quite enough,
And Mauritania now can stuff
Its sand back whence it came in days of yore:

(Chorus)

Someone in the audience taped that song; I hear it produced occasionally even now, after more than thirty years. Long ago, but that trip was one of the happiest of my entire career.

I missed the next few eclipses for reasons which were quite clear-cut. On 27 June 1976 my mother had had her ninetieth birthday; she was on great form, and it was a great occasion. We threw a major party at the Selsey Hotel, and she enjoyed it immensely; she was still able to get about. But at last the years started to take their toll. Mentally she was fine, right up to the end, but physically she was failing, and there was absolutely nothing that would take me away from her side. 'Woody' – Mrs. Woodward – came originally as a housekeeper, but turned into the most devoted friend, and for the last two years of Mother's life she did day duty, which I took over at night. Everything else went by the board, all I did was to cope with *The Sky at Night* and work away at my typewriter when I could, snatching sleep during the day if possible. Mother knew the score: 'My body is worn out' she said, and of course she was right.

Though I knew it was going to happen, I simply could not take it when the curtain finally fell, on 7 January, 1981 – the worst day of my life since Lorna made her exit. But there it was, and the only course was to settle down and accept the situation. Woody stayed – until she too died more than ten years later, leaving an unfillable void.

Back to eclipses... The East Indies, in June 1983. I joined a party bound for Java, which was new territory for me. Shortly before eclipse day the skies were hopelessly overcast, and the outlook was bleak. However, we kept hoping; statues of Buddha were everywhere, and we offered up prayers on the grounds that since he was the local god he might feel able to help. He did; the clouds rolled away, and we all had a perfect view. One of our party commented that he was a useful old Buddha. (I *think* he said Buddha.)

Our next eclipse foray, in 1988, took us to Talikud Island. I had never heard of it, but when I looked it up I found that it is small (a few miles long) and lies in Davao Bay in the Philippines, within six degrees of the equator. Davao is the main city of Mindanao, the southernmost of the larger Philippines – the only other major town there is General Santos City – and it is decidedly unstable politically, which is why many people steer clear of it.

Davao lay right on the centre of the track of the total solar eclipse of 17-18 March (the eclipse extended over midnight GMT). The Explorers Travel Club organised an expedition there, and I was invited to take part. This seemed an excellent idea, and so on 12 March I rendezvoused with the other members of the party at Gatwick, and boarded the plane for Manila.

We travelled by Philippine Airways. Once in the air we had no real problems, but timekeeping is not the airline's strong point, and delays of four or five hours are regarded as quite normal. However, we reached Manila eventually, and paid a visit to the local Planetarium to hear a lecture delivered by a speaker whose voice reminded me strongly of a duck in the distance. We then flew on to Davao.

The Filipinos do not believe in doing things by halves. Instead of one set of terrorists, they have three. Of these, the NPA (National

People's Army) is fairly conventional, and is modelled on the lines of the IRA. The MNLI is trying to liberate somebody from something, and there are also supporters of ex-President Ferdinand Marcos, who was overthrown a few years ago; at the time of the eclipse the current President was Mrs. Corazan Aquino. All these groups tend to blow things up. They also blow each other up, but on this occasion they were kind enough to put out a combined Press statement to the effect that since we were not involved in any Filipino affairs we were quite welcome, and they did not intend to blow us up. I thought that this was jolly sporting of them. They kept their word, and in the event, we heard gunfire only once.

When we reached Davao the temperature was of the order of a hundred in the shade. Davao itself is a fair-sized town – larger than General Santos City – and we had chosen it because we expected to find the best conditions there, but because any cloud was liable to hug the coastline we planned to go across to Talikud Island, with its palm-trees and glorious beaches. You can see it easily from Davao, and reach it by sailing across the bay in a rather flimsy outrigger boat.

The members of the party had various types of equipment, ranging from the elaborate (Bob Turner, Douglas Arnold and others) to the rudimentary (me, with my 400mm telephoto lens on a somewhat elderly Pentax camera). There were more than fifty of us all told. On 16 March we made a reconnaissance trip, and selected our sites. We decided upon the beach, so that the eclipsed Sun would (we hoped) show up over the palm trees, and the Moon's shadow would rush in toward us from the sea.

The real problem was the weather. Contrary to expectation, clouds had gathered, and on the morning of eclipse day the sky was overcast. Should we go to Talikud Island, as planned, or stay in Davao, or make for General Santos City? In the end most of us opted for Talikud, and at 5am Filipino time we set off. When we reached the island the clouds were still there, and we set up our equipment more in hope than in expectation.

First Contact – and still not a break in the clouds. The gloom increased. Totality was due at 9.07 Filipino time, but by half-past eight the situation was looking decidedly grim. It was worse by 8.45, when a gentle rain started to fall, and Bob Turner hastily covered his telescope with a large, multi-coloured umbrella which he had had the foresight to bring along.

I think that most of us had abandoned all, hope by 8.55. Little more than ten minutes to go; the light was fading, but still no break. I wondered what was the best course. Should I take my camera off its tripod, put in my wide-angle lens, and simply try for the Moon's shadow? Luckily, I decided against it; hope springs eternal. Then the rain stopped, and to my delight I saw the thin crescent of the Sun. Hastily I pointed my camera in what I hoped was the right direction, and took an exposure at 1/250 of a second (I was using professional 100 ISO film). The cloud was thinning, but another dark mass was approaching, and I offered up a prayer that it would not arrive before totality.

Abruptly the light level dropped. Glancing over my shoulder, I saw the Moon's shadow coming across the sea; it was more dramatic than anything I had expected. There was the flash of the Diamond Ring, and there, miraculously, was the eclipsed Sun.

It was amazing. A certain amount of light remained, and this drowned the outer corona, but the inner corona was bright, and there were two glorious red prominences. I suppose that the overall effect was enhanced by the idyllic scene below; the thin cloud enhanced the prominences at the expense of the corona. I doubt whether Venus or Jupiter could have been seen, but I had no time to look. Everything was still. Nature had suddenly called a halt.

We had three minutes of totality, and I took a series of exposures. Since I had had virtually no preparation time, all I could do was to bracket from about 1/10 second up to 4 seconds, and hope that at least one picture would be acceptable. There was no alternative, because it was impossible to gauge the effects of the residual cloud. Mercifully the dark mass kept away, menacing though it looked. The tension remained until Third Contact, with the flash of the Diamond Ring before light flooded back over Talikud Island. We had been incredibly lucky, and I did obtain one picture which I think is rather good. We said farewell to Talikud with considerable affection, and waved goodbye to the local residents – all four of them.

Subsequently, I found that I had had one extra piece of good fortune. Mrs. Aquino, President of the Philippines, had travelled to General Santos City to view the eclipse; I had been invited to join her (I have no idea why) and had been expected there, where the cloud cover was complete. Luckily I had not received the invitation, and so was able to send a perfectly genuine apology.

I was again lucky in 1991. The track of totality crossed Central America and also Hawaii; most people went to Hawaii, but I opted for the Mexican border, and was rewarded with a perfect view from the verandah of a luxury hotel, with a cloudless sky, a Sun nearly overhead and a glass of wine by my side – whereas conditions in Hawaii were at best mediocre.

For the eclipse of 24 October 1995 I was on a ship, the *Marco Polo* in the middle of the China Seas. It was a long totality (over six minutes) and we had a Norwegian captain who managed to get everything right; he manœuvred the *Marco Polo* into the one area which was free of clouds, and he aimed us in the right direction, so that we had an unobstructed view. With me was Chris Doherty, then aged seventeen. His father, Paul, was the famous astronomical artist, whom I had known since he was a boy. Tragically, he died of cancer five years ago. Chris is a photographic enthusiast, and had now taken his degree at Derby University. Since 1995 he has been my usual companion on 'away trips', and is generally taken for my son (which, frankly, from my point of view he might just as well be).

On this occasion he took some splendid pictures, as you can see. For the eclipse of 26 February 1998 we were again on the ocean wave, this time on the good ship *Stella Solaris,* not quite the equal of the *Marco Polo*, perhaps, but excellent all the same. This time we had a Greek captain, who again got everything right. The *Sky at Night* team had come along, with Pieter Morpurgo as producer, and we presented a full programme; my rôle was to do the commentary, so that I had no time to take photographs – I left that to Chris.

All went well (though I noted that one dear lady, clearly not an astronomer, sat in a deckchair and read a book throughout totality), and the programme was a success – it even won a prestigious award. I have one more happy memory. We were still on board on 4 March, my birthday, and a deck party was thrown for me, plus the gift of a fine Meade telescope. At the ship's concert I hope I did not disgrace myself on the xylophone…

After that we began making preparations for the eclipse of 11 August 1999, the first English totality since 1927 (note that I say English, not British, because the eclipse of 1954 was just total for a couple of seconds for the northernmost of the Shetland Isles). The 1999 track crossed the Scilly Isles and parts of Cornwall and Devon before going on into Europe. Of course the English weather is notoriously uncertain, but for *The Sky at Night* coverage we settled on Falmouth in Cornwall, though as a safeguard we arranged to receive pictures from a high-flying Hercules aircraft. With me and the television team at Falmouth were Peter Cattermole and Iain Nicolson.

The forthcoming eclipse generated a tremendous amount of interest all over the country, and the West made ready for traffic jams; it was even suggested that Cornwall might sink under the weight of the influx of visitors. I have one awful confession to make. On 1 April I went to Cornish radio and announced solemnly that the eclipse had been postponed for a month, and would now take place on 11 September instead of 11 August. I was followed by the Cornish traffic controller, who said that he would now have to rearrange his schedules. We didn't think that people would take this seriously on April Fool's Day, but we were wrong – the phone lines were jammed, and the Cornish and Devon County Councils went bananas! Echoes of that lingered on well into the summer...

Near the date of the eclipse I made a good many broadcasts to stress the danger of looking directly at the Sun through any optical equipment. In previous eclipses there had been many cases of eyesight damage, and I wanted to make sure that nothing of the sort happened again. I was fairly successful, though a misleading Press release from the Ministry of Health did not help, and advised against looking at the eclipse at all (!). Peter developed a very suitable device; fit a tiny telescope or pair of binoculars to the inside of a large cardboard tube, aim at the Sun, and project the image on to the bottom of the tube. Quite safe, and refreshingly easy to make. Even I could do it.

Morning of 11 August... And at Falmouth, total cloud cover. Despite this, we set up our cameras and instruments in the appropriate positions, watched by a crowd of people who seemed to materialize from nowhere. About twenty minutes before totality we had a brief glimpse of the crescent Sun through the cloud, but that was all, and we had to resign ourselves to defeat, though the pictures from the Hercules were good. Rain began to fall, and we crouched under umbrellas, muttering things such as 'Tut, tut!' and 'Fiddle, fiddle, fiddle!'

Actually the onset of totality was quite dramatic – an eerie gloom, quite unlike normal darkness; even the seagulls fell silent. Many people decided to take pictures, and the scene was lit up by photograph flashes. Then the darkness lifted, and Nature woke up. Sadly we folded our umbrellas, wrapped up the television broadcast and retreated to our temporary studio, to lick our wounds with the aid of several bottles of Irish whiskey.

We were not alone in our disappointment; the British Astronomical Association party had been clouded out, and most of the track of totality was overcast, though there were a few clear areas – the Scillies, Newquay, the Lizard, and Alderney in the Channel Islands. However, the partial phase was seen over much of Britain, and was widely observed.

The next English totality will be in 2090, again over the West Country. I can only hope that those who go there will have better luck than we did in 1999!

15 The Grand Tour

I always think of the 1970s and 1980s as the Years of the Planets. The Moon was relegated to a secondary rôle after Apollo 17, and attention shifted elsewhere, culminating in the Grand Tour of Voyager 2. The main attention also shifted away from Europe. The Russians sent some successful spacecraft to Venus, but otherwise they had no luck at all.

My own rôle was, to put it mildly, minor. From a purely observational point of view I was busy monitoring Mars, Saturn and particularly Jupiter, mainly from Selsey, but otherwise my only contribution was to make regular reports on television – not only on *The Sky at Night* but also from NASA headquarters during the actual missions. I flew to and fro across the Pond, because the BBC wanted me to be 'on site' at the critical moments, and frankly I would have hated to have been anywhere else.

Looking back, it amazes me to realize how much we did *not* know about the planets before the rockets started to fly. When I first began to take an active interest, which was a long time ago (my original observing books date back to 1933), many astronomers still believed in the canals of Mars, built by a brilliant civilization far in advance of ours. In particular, the Martians were assumed to have more sense than to slaughter each other, as we Earthmen have been doing ever since we emerged from caves. I have always been impressed by the wise words of Percival Lowell, who was admittedly wrong about Mars but right about many other things:

'War is a survival among us from savage times, and now affects now chiefly the boyish and unthinking element in the nation. The wisest realize that there are better ways for practising heroism and other and more certain ends of insuring the survival of the fittest. It is something a people outgrow. But whether they consciously practise peace or not, nature in its evolution eventually practises it for them, and after enough of the inhabitants of a globe have killed each other off, the remainder must find it more advantageous to work together for the common good.'

Lowell wrote that comment in 1909. Since then we seem to have regressed rather than advanced, and at the present moment (January 2003) there are at least a dozen wars going on. I bitterly regret that there are no Martians; if there were, I would be the first to invite them here and ask them to show us how to run our own planet.

I have already said something about Mariner 9, the first probe to go into orbit around Mars and show us the giant volcanoes. At about the same time (December 1971) the Russians did land a spacecraft there, Mars 3, but after arrival the transmissions lasted for only a minute or two before all contact was lost. I was broadcasting on the Overseas Service, and the interviewer asked me, quite seriously, whether I thought that some irritated Martian had gone up to the probe and switched it off. I said that in my view this was unlikely, but of course I could not definitely rule it out. We would have to wait for further information.

It came with the Vikings, which were launched in 1975 and reached their target in the summer of 1976. Each Viking – there were two of them – consisted of an orbiter and a lander. Each was successful, and we obtained the first photographs direct from the Martian surface, showing a striking absence of canals but a vast number of rocks strewn across the landscape, under a yellowish-pink sky. But the orbiting sections sent back the first results, and one image showed a rock which looked uncannily like a humanoid face. It lay in one of the ochre 'deserts', Cydonia, and of course it attracted a great deal of attention, with all the cranks buzzing around. There were immediate suggestions that it was artificial, carved by the local inhabitants for reasons of their own – a sort of Martian sphinx, in fact. This led on to the usual stories about lost civilisations, abandoned cities and the like. Before long we were back to claims that the Martians were still alive and kicking, and had even sent representatives to Earth to keep a watchful eye on us.

I well remember one interview soon after the Viking picture was published. It was destined for a local radio station in the Midlands, but I doubt whether it was ever transmitted; if it was, it must have been heavily censored. The interviewer was a brash young man aged, I should think, about twenty. I still have the tape, and part of it sums up the general theme:

Interviewer. Do you believe that the Face is artificial?

Me. No, I don't.

Interviewer. Why not?

Me. It's clearly a natural rock formation – an ordinary clump of material. There's nothing non-natural about it.

Interviewer. Nothing non-natural? You (*expletive deleted*) well know that that isn't true. You can see the eyes and the mouth.

Me. Pure chance light-and-shade effects.

Interviewer. That's what all you (*e.d.*) experts say. You're (*e.d.*) trying to cover up what's been discovered – a Martian race.

Me. I fear not. There's no advanced life on Mars.

Interviewer. There (*e.d.*) well is, and you (*e.d.*) well know it.

There didn't seem much point in prolonging that interview, because I knew full well that nothing I could say would have the slightest effect. As soon as I hear hints of a 'conspiracy theory', I realize that the only thing to do is to give up. Even in 1998, when a new probe, Mars Global Surveyor, sent back images taken from a different angle, showing that what had been taken for eyes, nose and lips were nothing more than hills and ridges, the conspiracy theorists remained convinced. They also fastened upon a small crater, Galle, which has a curved ridge on its floor which made it look like a 'happy, smiling face'.

Yet even though there are no Martians, we cannot yet be sure that there is no life there at all. If it exists, it must be very lowly. Yet it would be carbon-based as is all life on Earth, and its discovery would be of tremendous importance even if it were no more advanced than an amœba. Before long, we ought to know.

As the lander of Viking 1 was descending toward Mars, on 19 June 1976, I was in BBC Studio 7 with Garry Hunt and Geoffrey Eglinton, both of whom were closely involved with the whole Viking programme. We had some models of the spacecraft, and we experimented by dropping them on to a representation of the Martian surface. I recalled a comment by Garry: 'At least we've proved one thing. If you drop them too hard, they break.' Actually the touchdown speed of Viking 1 was less than 7mph, and all went well. The lander came down within 26 feet of a large boulder which would have brought the mission to an abrupt end if there had been a direct hit.

Jupiter, the Giant Planet, was also on NASA's menu, and during the 1970s four probes were sent to it, two Pioneers and two Voyagers. All performed excellently. For all the passes I was at Pasadena, NASA's headquarters, and it was a fascinating time. To me, the most impressive room in the whole complex is the DSN, or Deep Space Network, where all the main planetary probes are monitored continuously. Few visitors to the DSN are allowed, but I was one, and I remember thinking that it might have come straight out of Quartermass. Even now, in 2003, the DSN is still in touch with Pioneer 10, which by-passed Jupiter on 3 December 1973 and is now more than 4,000 million miles from the Earth. Its signals are faint indeed, but they are still detectable. Yet we cannot hope to monitor them for much longer, and what will eventually happen to Pioneer 10 we know not. One day, perhaps some remote civilization far across the Galaxy will salvage it and remove it to a cosmic museum…

My own real research in connection with Jupiter had ended some years earlier. Astronomical bodies send out radiations over the whole range of wavelengths, and in the 1950s it was found – by accident, it has to be admitted – that Jupiter is a radio emitter. It was important to decide whether these radio waves came from the whole body of the planet, or whether they issued from discrete features such as the Great Red Spot, a whirling storm which has been seen on and off (more on than off) ever since the seventeenth century, when telescopes were invented and became powerful enough to show the planets in reasonable detail. The Jupiter Section of the British Astronomical Association, of which I was a member, collaborated with the radio astronomers; our rôle was to check the surface features and see if we could link them with radio bursts. Despite its great size (diameter nearly 90,000 miles) Jupiter spins very quickly; a Jovian 'day' is less than ten hours long, and the rotation period at the equator is appreciably shorter than it is at higher latitudes, because Jupiter does not spin in the way that a rigid, rocky globe would do. The trick is to watch the surface features, and note the times when they reach the central meridian, i.e. the centre longitude of the disk seen from Earth. For example, take the Great Red Spot. If there was enhanced radio emission each time the Spot crossed the central meridian, we would have proved our point.

Whenever Jupiter was well placed, as it is for several months in every year, I was in regular touch with two particular colleagues, Theodore Phillips (a clergyman) and Ben Burrell (by profession a railway porter). We had similar-sized telescopes, and we always agreed well. In the end we did give definite proof: there was no radio link with discrete surface features. However, there is a strong link between the radio emissions and the position of Jupiter's volcanic satellite, Io – an association first pointed out by K. E. Bigg, who is not an astronomer at all, but a meteorologist.

I remember one interview I gave from Pasadena during the first Jupiter pass. I knew the interviewer, and I was fairly sure what line he would take. 'Why waste money on space research, when there is so much to be done down here?' 'What is the use of it?' 'How many hospitals could you build with that money?' etc., etc. So as soon as we went on the air (live), I took the initiative. 'Nice to talk to you,' I said. 'It's all going well, and up to now we haven't even seen the "weary willies" who try to play down space research and say that we ought not to spend any money on it. You know the sort of people – they are too stupid to realize that space research is linked with all other sciences, including medical science. They simply haven't done their homework, and they are a nuisance; in earlier times they would have objected to the development of the wheel. As soon as I hear the question "What's the use of space research?" I know that I'm dealing with an idiot. Of course, I know that you would never ask a question as stupid as that.' As it was exactly the question he *was* going to ask, he was decidedly nonplussed, and I controlled the interview with no difficulty at all. Some time later we met in a London studio, and he recalled the incident, even referring to me in terms which inferred that my parents had met only once, briefly, at a masked ball. In compensation, I bought him a large gin and tonic!

I was back at Pasadena for the Voyager encounters with the outer planets, Saturn (1980 and 1981), Uranus (1986) and Neptune (1989) – now purely as a BBC reporter, not a researcher in any sense of the term. The two Saturn passes, one by Voyager 1 and the other by Voyager 2, produced stunning pictures, both of the planet itself and its satellites, even though Titan did hide coyly behind a veil of murky smog. Then came encounters by Voyager 2 only; once past Saturn, Voyager 1 went on its way, speeding out of the Solar System to wander among the stars.

Uranus, more than 1,780 million miles from the Sun, is an

oddity. It has less than half Saturn's diameter (31,000 miles, as against Saturn's 75,000), and is a different sort of world, made up largely of 'ices', and with at best only a negligible inner heat source. Its 'year' is 84 times as long as ours, its 'day' amounts to 17¼ hours, and it has five principal satellites, all smaller than our Moon. No Earth-based telescope will show much on its pale greenish disk. Of course it has a gaseous surface, with an extensive atmosphere but probably only a relatively small solid core.

One of the most curious things about it is the tilt of its axis. The Earth's axis is tilted by 23½ degrees to the perpendicular to the orbit, which is why we have our seasons; with Uranus the tilt is 98 degrees, so that it 'rolls along' as it moves round the Sun, and there are times when one of the poles is turned directly sunward, so that the pole is then warmer than the equator. But which is Uranus' north pole, and which is the south? Trying to clarify that on television was quite daunting.

The plane of the orbit is known as the ecliptic, and our north pole lies above the galactic plane. The plane of Uranus' orbit – that is to say, the Uranian ecliptic – lies within one degree of ours. The International Astronomical Union, whose word is law, has decreed that with other planets all poles that lie above their ecliptic are south poles. When Voyager 2 passed Uranus, it flew above the sunlit pole – and the seasons there are very strange, first one pole, then the other, has a 'night' lasting for 21 Earth years, with a corresponding 'midnight sun' at the opposite pole. On this basis, it was the south pole which was surveyed by Voyager. However, the Voyager scientific team reversed this, and referred to the sunlit pole as the *north* pole.

Is that clear? I hope it is clearer than it seemed to be to the viewers when I made my broadcast!

When we had finished our coverage of the Uranus mission, and Voyager 2 was en route for its next and final target – Neptune – I set off for home and a BBC studio. I flew across the Atlantic, landed at Heathrow, and was met by the Press. 'What do you feel about the *Challenger* disaster?'

The Shuttle tragedy had taken place while I was in mid-air, so that I had to be brief before I could make any comment at all. You probably know what had happened. The space shuttle *Challenger*, carrying a full crew – including a schoolteacher – had exploded soon after launch, killing all the astronauts on board. I was utterly taken aback. What was there to say? I did not know any of the *Challenger* crew personally, though of course that made no difference. I said what I could.

It was a salutary lesson to everyone. Space is a dangerous environment, and will always be so; anyone venturing there is running risks. Before *Challenger*, there had been so many successes that some people were starting to become blasé. It is true that more lives were lost during the pioneer days of aeronautics than in the earliest period of astronautics, but danger was never far off, and in the case of *Challenger* human error was to blame. There was a potential fault in the launcher, and the mission ought to have been postponed until exhaustive checks had been carried out, but the political planners were anxious not to delay, because of the presence of the schoolteacher. It was a propaganda flight, and NASA paid a terrible price.

From Uranus, Voyager 2 made its way out of Neptune, more than 2700 million miles from Earth, so that a radio signal would take over four hours to each us. The signal strength was incredibly low, and radio telescopes from elsewhere were brought in to help, notably those at Goldstone in California, Robledo in Spain, Parkes in Australia, and the Very Large Array in New Mexico. It was a truly international effort. Eventually, after a course correction in August which altered Voyager's speed by 2.1mph, the probe reached its target within six minutes of the scheduled time.

When I reached JPL (Jet Propulsion Laboratory), on 10 August 1989, I found an air of great excitement. We all knew that this was potentially the most fascinating encounter of all, and also that it would be the last for many years. All ideas of sending Voyager on to the ninth planet, Pluto, had to be abandoned when the mathematicians found that it would have meant burrowing into Neptune's globe, which did not seem to be a good idea. So we were seeing the end of an era.

Officially I was purely a BBC reporter, and I was scheduled to present daily programmes, with the support of Garry Hunt (who was a tower of strength throughout all the missions, and was, of course, one of NASA's Principal Scientific investigators). But I did sometimes have to revert. The whole mission was being extensively covered by CBC, the main Canadian TV network, which had a regular audience of many millions. Just as Voyager was closing in to Neptune, the CBC producer came up to me. 'We have a problem. The astronomer who was going to talk about Neptune's atmosphere has been called away, and there's nobody to fill the slot'. He paused. 'You're an astronomer. Can you do it?'

'Of course I can. How long do you want?'

'Ten minutes, this morning. Does that give you time to write a script? We're due on in half an hour.'

I looked at him pityingly. 'Why on earth should I write a script? If you want ten minutes, to camera, that's fine.'

The producer looked dubious; clearly he wasn't used to a TV broadcaster speaking 'off the cuff', but in the end he put his trust in me, and at the appointed time I did the slot; I came out after 9 minutes 47 seconds. The producer beamed. 'What's your fee for being one of our regular team?'

'Nil.' I said. I was on the spot; the BBC had no objection; why should I not help? Evidently I was an oddity by transatlantic standards, but in the end I made quite a number of broadcasts on the CBC network – subsequently I was delighted to find a large crate of Canadian whisky in my hotel room!

The encounter did turn out to be enthralling. Neptune produced a Great Dark Spot, a quicker-rotating spot nicknamed the 'Scooter', and much else; it was a much more dynamic world than the rather bland Uranus. The pièce de résistance was the one large satellite, Triton, which was found to be partially covered with snow – not our kind of snow, but nitrogen snow. Triton is so cold that nitrogen, which we breathe in as a gas (it makes up 78 per cent of the Earth's atmosphere) freezes out. Moreover, there were nitrogen geysers spouting from below. All in all, I think that Triton was the most surprising world we have yet found.

Within a few days the encounter was over; by 29 August Voyager was already more than four and a half thousand miles beyond Neptune, and the planet and its satellite were little more than specks in the distance. Our farewell party at JPL was a mixture of emotions; elation because everything had gone so well, and regret because it was clear that we would never again meet in this way. Laurence Soderblom, one of NASA's leading scientists, summed it up very appropriately: 'Wow! What a way to leave the Solar System!'

There were other missions (for example Mariner 10 to Mercury), but it has to be said that after Voyager 2 there was something of a lull, and the planetary programme was not really re-galvanized until 1990, when America's space-craft Magellan was put into orbit round Venus and gave us our first reliable maps of that peculiar planet. And near the end of the century came new forays to the Moon and Mars, plus several asteroids. But before discussing these developments, I feel that I must say a little about the international aspect of astronomy, because I did make a contribution – albeit a very minor one.

16 O Argentina

When the Cold War was just hotting up (if you see what I mean) I had to go to the United States for a conference. I was there as a Moon-mapper, and I was delighted to be invited, but this was before I became associated with NASA on a more or less official level, and there was one obstacle to be overcome. I had just been over to the USSR, again as a Moon-mapper, and my passport had a Soviet stamp on it; this was not likely to go down well with the US immigration authorities, who are much the same as immigration authorities in all other countries (except Norway and Iceland). Therefore, it was prudent to lose my passport, and obtain a temporary one which did not have a Soviet stamp. There was no problem.

I mention this because although there is no doubt that relations between Washington and Moscow were strained, nothing of the sort affected the International Astronomical Union, which was – and is – the supreme authority in world astronomy. There have never been any Iron Curtains, and at all times the Russians, the Americans and everyone else have worked together. This did not apply to space research, because a vehicle which can launch a probe to Mars can also launch a nuclear bomb, and for some time China kept aloof. During Chairman Mao's Cultural Revolution the Chinese shot all their astronomers, which seemed to be rather extreme even at the time, and which they subsequently admitted had been a mistake. There was also the problem that Taiwan was a member, and Beijing did not approve. It was only in the late 1990s that all these little disputes were finally resolved.

The IAU is a purely professional body, and there are very few amateur members. General Assemblies are held every three years, in different countries. I first attended an Assembly in the 1960s, in Moscow, and it was most interesting – not only for the astronomy. The official languages of the Assembly were English and French (technically this is still true today, though in fact you can forget the French, which does not please our Gallic friends at all). The IAU rules lay down that if the speaker presents his paper in any other language there is simultaneous translation in the headphones provided. At Moscow, one paper was delivered in what I took to be Russian, and since my knowledge of Russian is limited to 'Nyet!' and 'Spasse-bo' I tuned in to the English translation. Nothing came through. I tried French, which was the same result, and also German, which I know would have been of no use to me. I then gave up. It later transpired that the English thought he was talking Russian; the Russians thought he was talking English (which he was), and the French and the Germans didn't know. As there was no Press release or preprint, and no illustrations, we waited for the paper to appear in the Proceedings. It never did, so to this day nobody has the slightest idea of what that paper was about. The audience applauded politely when he had finished. (For Heaven's sake, why doesn't everybody talk English? It is so much easier, and there is none of this ridiculous nonsense about genders!)

The IAU is in some respects a curious organization. You can't apply for membership; you have to be proposed, and seconded by three other members. Your name is then displayed at the next Assembly, so that your election can be confirmed or otherwise. I was proposed for membership in 1966, when the General Assembly was held in Prague, in what was then Czechoslovakia. I drove there with a friend of long standing, Dr. Gilbert Fielder, one of the world's leading experts on all matters connected with the

Moon. We took Gilbert's car across the Channel, and then made our way across Europe, which was then a rather safer place than it is now. When we reached Czechoslovakia we became hopelessly lost, but eventually we located Prague, and booked into our hotel.

Even though the Russians had not yet taken control in Czechoslovakia, the atmosphere was decidedly tense, though this did not affect us directly, and the Czech astronomers were as friendly as they could be. All the same, I was somewhat wary, and on my last night in the hotel I cast around for a 'bug'. I found it; it ran off the radio. I do know about these things so I recited 'The Walrus and the Carpenter' into it and left it. I imagine that the OGPU spent some time in trying to decode it, but by that time I was safely back in the West.

The first general meeting was held in the old Palace of the Kings – a truly magnificent place, and we were given a welcoming banquet. As we arrived, the Imperial Guard band was playing the lovely waltz *Baletky*, by Julius Fučik – the 'Czech Strauss', who is in my view almost the equal of the great Johann. It was most impressive, and if I hear *Baletky* today it takes me back to that magic moment.

Our hotel was equipped with those strange continental lifts which are doorless, and in constant motion, so that you simply step in when the lift arrives. You then go over the top, and come down on the opposite side. On one occasion we stayed inside, and some Americans were staggered to see us descend upside-down; it took some practice to enable me to stand on my head. I also have to admit that Gilbert and I were once seen sitting at a coffee table outside the main café, playing blow football with an ant. I forget who won...

On a more serious note, I recall a conversation with Professor Harold Urey, a Nobel Laureate and one of the world's great scientists. We were standing on a large map of the Moon which had been laid down on the floor of one of the lecture rooms, and he was trying to convert me to his theory that the lunar 'seas' had once been filled with water. I was highly sceptical, and we are now sure that there have never been any water oceans on the Moon, so that for once Urey was wrong and I was right – a situation that didn't happen very often.

One inevitable feature of General Assemblies is that there are many specialized meetings going on at the same time, so that one has to be selective. Of course I was concerned with the lunar and planetary sections, but I did go to one discussion which was outside my field. It was concerned with quasars, which had not long been identified, and which were known to be very remote and super-luminous; a single quasar can shine as brightly as a hundred galaxies, and a galaxy may contain a hundred thousand million stars. Since I am no cosmologist, and my mathematics are of schoolboy standard, I felt rather out of place, but I made one point. Is it possible that quasars are much dimmer than is usually thought, and are minor features associated with galaxies? If so, it just might be worth searching the areas round galaxies (particularly radio-emitting galaxies) to see whether there were any concentration of objects with quasar-like spectra. The idea was dismissed, but today there are some astronomers, such as Dr. Halton ('Chip') Arp, who take it very seriously. So did the late Fred Hoyle, who could often be right when everyone else was wrong (though equally often he could be wrong when everyone else was right). The idea of dim quasars is unpopular with most researchers, possibly because if it is correct the result will be that many PhD theses have to be consigned to the nearest waste-paper basket.

There was one more item in which I was involved, and which, sadly, turned out to be rather a failure. This was the setting-up of an International Union of Amateur Astronomers.

Astronomy is still just about the only science in which the amateur can be really useful. I suppose there are still chances for good amateur geologists and naturalists, but only in astronomy is there no doubt that amateurs and professionals work really well together. After all, the average professional spends very little time in actually looking at the sky; he uses electronic equipment together with powerful telescopes, and I am quite sure that there are many eminent professionals who could not identify any of the constellations apart possibly from the Great Bear. (I once had a phone call from a well-known cosmologist who told me that he had just found a bright nova, or new star. In fact he had made a completely independent discovery of the planet Saturn.) Of course, there are amateur societies in every civilized country, and there is good co-operation; for example in England we have the British Astronomical Association, amateur with a sprinkling of professionals, and the Royal Astronomical Society, professional with a sprinkling of amateurs, whose main offices are in the same building (Burlington House, Piccadilly). The amateurs and the professionals had never been brought together under one roof, so to speak, and it was suggested that this might be a possibility. So I was given the task of seeing whether an IUAA was viable.

I called a meeting during one of the less crowded mornings in Prague, and it was well attended, with representatives from most countries. I felt encouraged. I had prepared an opening address, and then asked for comments. I was invited to become provisional Chairman, and we formed a committee to work out procedures. A second meeting was equally encouraging, and the IUAA came officially into being. The outlook seemed bright, but I have to admit that things did not go according to plan, and to a large extent I blame myself. I was asked to become Chairman for the first year of the IUAA's existence; we established a Journal, we planned meetings modelled on those of the IAU, and we produced a membership list, but when the time came for elections I stood down, because the last thing I wanted was to be accused of trying to make the IUAA a personal venture. Arrangements were taken over by the delegates from Italy and the Irish Republic, and to be candid this was not an ideal combination. Over the next couple of decades there were sporadic meetings and issues of the

Journal, but the 'spark' was never there, and by now (2003) the IUAA has to all intents and purposes ceased to function.

Why did it go wrong? First, it was too ambitious. My initial idea was to have a Journal which merely acted as an exchange of information, not to produce papers which would clash with those of national societies. Also, we tried to make it multilingual; we ought to have followed the lead of the IAU and kept to English. So far as meetings are concerned, amateurs have the disadvantage of not being sponsored, so that if we decided to call an Assembly in, say, Sweden – as we did in 1983 – only a minority of members could afford to go.

It was a disappointment, and despite all the problems we ought to have done better. I hope that in the future the IUAA will be revived, and that those who are prepared to organize it will not make the same mistakes as we did.

I missed the next IAU General Assembly, in America, but the 1971 Assembly was held in Brighton, right on my doorstep. I did present a short paper, dealing with some aspects of lunar observation, but my main task was in coping with the media, gently pointing out to reporters and interviewers that there really is quite a difference between astronomy and astrology. There are times when I despair of modern education.

During the rest of the 1970s I did very little travelling, apart from a couple of brief eclipse trips, because my mother was entering her nineties, and although she was mentally 100 per cent we both knew that she could not go on for ever, and I had no wish to be away from her. So I missed out, and it was 1986 before I went to my next Assembly, held at the University of Patras in Greece. It was great fun. The Greeks are splendid people (in my view, they and the Scandinavians are much the nicest of the Europeans) and it is impossible to become annoyed with them when things go awry – as they usually do; a bus strike did not help. With John Mason and others I flew to Athens, managed to get a train to Patras and booked in at an hotel in the main street. This turned out to be a wise choice. Each evening after the end of the official sessions the café chairs on the pavement outside the hotels would be occupied, and a vast amount of retsina was consumed. (Either you love retsina, or you hate it; I think it is the most attractive of all wines.) Generally we were the first to open the proceedings, followed closely by the *Sky and Telescope* team from the United States. Frankly, at IAU Assemblies the 'after hours' activities are always among the most interesting and profitable.

We were also treated to a performance of a Greek tragedy. As my knowledge of Greek, ancient or modern, is limited to the alphabet, the finer points of the plot eluded me, and I am ashamed to say that some of our party were inclined to giggle. Luckily our Greek hosts didn't mind.

The 1988 Assembly was held in Baltimore, home of the offices of the Hubble Space Telescope. Then, in 1991, the venue was Buenos Aires in Argentina, the first time an Assembly had been to South America. And thereby hangs a tale.

At each General Assembly there is an IAU Newspaper. It goes out every morning to every delegate, covering the previous day's papers as well as announcements, articles and information about the events for the day. Normally it is edited by a professional astronomer with a team of reporters and translators, a suite of offices, and rows of gleaming computers. (Dr. Steven Maran had been Editor at Baltimore). This time my old friend Dr. Derek McNally, the IAU Secretary-General, asked me to be Editor. I have an unworthy suspicion that he knew what I was in for, and I was under no delusions. I co-opted Dr. John Mason, who at that time was President of the British Astronomical Association, and we packed our bags.

The Assembly took place immediately after the total solar eclipse that I described earlier; I went to Mexico and John to Hawaii. After the eclipse John joined me, and we set off for Buenos Aires ten days before the Assembly was due to start, because I wanted to have sufficient time to sort things out; I had an uneasy feeling that all would not be well. At least we made a firm resolution: 'Don't mention the Falklands!'

In many ways it was a curious situation. I am sure I am the only amateur ever to have held an official position in the IAU, and the Newspaper was simply handed over to me; John, whom I had calmly appointed Assistant Editor, has a string of degrees (more than most thermometers) but was not then even a member of the IAU. My first task was to have him elected. Rather than to propose him myself, I arranged for him to be proposed by one Astronomer Royal and seconded by another. It worked like a charm.

Why Argentina? Partly because it was new territory so far as the IAU was concerned, but also because Argentina has quite a reputation in astronomy; there is a major observatory at Cordoba, founded well over a century ago, and another observatory at La Plata. The Argentines were keen to have us, and the stage was set.

On our return to London, I gave a paper to the British Astronomical Association entitled 'How to Produce the IAU Paper'. It was greeted with raucous laughter. It was not published – I vetoed it – and I am not quite sure that I ought to publish it even now, but after a lapse of more than ten years I doubt whether anyone will object. So here it is, exactly as I delivered it. I can assure you that every word is true, and absolutely nothing has been either invented or exaggerated. In fact, there may well be episodes which I have left out. Here, then, is the paper:

> I rendezvoused with John Mason as arranged, and our journey by plane was uneventful. I will only add that at Los Angeles, before departing for Argentina, we stayed at an hotel where there was a billiard table. We had half a dozen games of snooker, demonstrating that for sheer, utter incompetence we were dead level.
>
> I had booked rooms for us in Buenos Aires. On arrival, we found that in our hotel we were on the 18th floor. The hotel was under renovation, and there was only one small lift,

which could hold five people (four, if I were one). We wanted to go to our rooms, and it took us twenty minutes before we could cram into a lift. This was clearly not 'on', and so we asked to be moved down. This was naturally impossible. We therefore sought, and did find, another hotel, the Savoy, where we settled in. We did have one night at the first hotel, however. In each of our bedrooms the toilet flooded. The radiators would not turn off, so that the temperature in each bedroom resembled that on the surface of Venus. When I did manage to turn off the radiator, the handle broke off. We departed with relief; on arrival at the Savoy the toilet in my room flooded, but I managed to get it fixed before too long.

Things seemed to be a little amiss. The Argentines had set up a Local Organizing Committee (LOC), whose organization did not extend to finding the Editor either officers, helpers or even a computer. We were scheduled to produce the newspaper by desk-top publishing. Obviously we had to have a computer, and we managed to obtain just one, which has no printer. We then found that the newspaper, which we called *Cruz del Sur* (Southern Cross) was to be printed in La Plata, forty miles away.

When we obtained a printer for our tiny computer, it had no plugs. We had to use a hotel bedroom (John's) as an office, because the offices in the San Martin Centre were not ready, and in any case we had not been allotted one. We managed to plug the computer into the socket in the wall, so that with live wires and a live transmitter on the floor it was absolutely lethal. The printer only just fitted. It was not compatible with the IAU computer, and neither was compatible with the printer's computer in La Plata.

We were able to meet our one helper, who spoke no English. I speak French, but no Spanish. Neither did John. We managed initially by semaphore. (I must admit that he turned out to be very useful indeed.)

Having set up an office in the bedroom, I called for a typewriter, because I prefer to write and edit on a manual machine rather than a computer, to which I am quite unused. The LOC sent up a typewriter, which was totally static and refused to work at all. Having selected the typeface for the newspaper, we decided to go to the bar of our first hotel (our new one had no bar) for a drink. The Argentine currency is the australe – 10,000 to the dollar. We went into the bar with half a million australes. It wasn't enough for two rounds of drinks. Of course, the hotel did not take cards or travellers' cheques, and it took an hour to foist a traveller's cheque on them.

Having set Issue No. 1, more or less, we waited for the opening of the General Assembly on the Monday. We commandeered part of a room on the first floor, and started in. The Press Office had no phone, and no fax. We did eventually get a phone, but it would have been impossible to ring out of Argentina even if the phone worked (which, for most of the time it didn't). Eventually, one phone and one fax was shared by all 1,200 delegates. Moreover, nobody could find out the phone number.

We naturally wanted some IAU notepaper. None could be found. There was however a photocopier, so we cut off the top of one of the IAU letters and photocopied it on to ordinary paper; this became our official stationery. An official photographer had been assigned to us, but he refused to take any pictures unless he was paid more money. In the end I took most of the pictures myself, and had them developed locally.

In trying to organize something in Buenos Aires, I had my first car crash. The Argentine traffic was officially described as 'aggressive and chaotic'. It is; there are no rules. Moreover, the pedestrian lights and the traffic lights do not synchronize, so that both are green together, and the result is rather hair-raising, particularly with Argentine taxi-drivers. Also the roads were under repair, and workmen were literally using spades and hammers. The fact that the traffic lights did not synchronize made no difference, because no traffic took a blind bit of notice of them anyway.

The lifts in our new hotel were now faulty, and on the first evening I was stuck in one for twenty minutes.

Each delegate was issued with a badge. Mine, of course, was an official one. On it was inscribed *Patrick Moore, Français*. I took it off and replaced it with another: 'Je ne suis pas Frog'.

I acquired a second typewriter, which immediately came to bits. We mended it with sellotape, and it just about functioned, though the letter q was liable to fly off. As I had to do some of the keying-in, I had to learn computer setting, and I felt totally alien. At this point the computer in the main IAU office broke down. The IAU hierarchy wanted to borrow ours, but (a) we couldn't spare it, and (b) it wasn't compatible anyway.

The phone in the so-called Press office was now installed, but it couldn't receive incoming calls.

There were no mail boxes for any delegates. These would not be ready until the Wednesday, halfway through the first week of the General Assembly. Outside the San Martin Centre, workmen were now digging up the roadway with hammer and chisel.

The IAU phone then broke down completely. The officers began to use ours, which also broke down. Our Spanish helper was then stuck in the lift for half an hour.

The mail boxes arrived. Alphabetically they were chaotic. My number was 5003, in between Nos. 462 and 541. Jean-Claude Pecker (Paris Observatory) suggested that the LOC had arranged the boxes according to the third letter of the delegates name, so mine would be under O.

There was no table for distribution of the newspaper. The Argentines said that if the newspapers were laid out, delegates might take more than one copy each, so they made it impossible for anyone to get any. We then took over the distribution ourselves.

I then sat on my monocle. John mended it with a paper-clip.

The mailboxes were moved; my number 5003 was now between 6 and 24, to the upper right of the main mailbox block.

The meeting rooms were switched. Room K + L became Room L, since Room K had been wrongly signposted, and delegates trying to get to a meeting there found themselves joining an Argentine drama class. In Room L, there had to be a special attendant to switch the lights on and off, since every time this was done large sparks came from the transformers. In Room M, the overhead projector was so scratched that anything written on it was completely illegible.

On Friday the mail boxes were moved again. I was now between 172 and 49. I may add that since we had no official reporters, we had to conscript people, and of course all the written contributions had to be heavily sub-edited, as they were in curious English. In the end I wrote most of them myself.

The next session was opened by Mr. Menem, the President of Argentina. He insisted that on the platform there had to be an odd number of people, so that he could have the central position – otherwise he wouldn't come; as there had been six originally, one of them had to be shifted off the platform. Mr. Menem then gave an excellent address, ending by saying that it was a great honour to have the IAU meeting in Argentina, and he was well aware of the great importance of astrology.

The phone cable in the Press room was now coiled round the computer. We freed it, but the phone was then useless.

A mysterious third bed appeared in John's hotel bedroom. Mine had the conventional two; we never found out why John's had three.

The mailboxes were moved again. Mine could not now be found.

It was then Sunday. To prepare the next issue we went to the San Martin Centre, where the office was. It was locked. Nobody had the key, so we retired to my bedroom, minus the computer.

On Monday we were back in the office. We had a lot of material for the Tuesday issue on the computer, but an electrician in the outer corridor, working on some repairs to the wiring (which was Roman), pulled a plug, wiping out two hours' work on the computer which we had set. Our Spanish helper dashed outside, and said things which in Hispanic presumably meant 'Dear Me!' Naughty, naughty! Please don't do that again!'

On Monday the times of the meetings were wrongly given by four hours. We were besieged by people who wanted to know what was happening where.

The mailboxes were moved again. I found mine, between numbers 188 and 1079.

The IAU Resolution had to be printed in the last issue. I asked how many pages would be required. The estimate given to me proved to be wrong by a factor of 10. I wanted to take the Resolutions out of the IAU computer, so save re-setting, but it was of course incompatible with ours. However, John found an old disk which had been left behind by a visiting professor, and by some inspired juggling announced that it could be made to fit. By then it was past midnight, and all the IAU staff had left except Monique, the French IAU Secretary, who referred despairingly to 'ze mess we are all in'. ('Mess' wasn't her actual word, but that was what she meant.)

It was now coming up to the penultimate day. We left the San Martin at 2am, and walked back to the hotel. At 7am the next morning we walked back, ready to begin the day's work. Smoke and flames were issuing from ahead. 'I hope that isn't the San Martin on fire,' one of us said. But it was. When we arrived dense smoke was billowing out, and delegates and officials were milling around like ants round an ant heap. The Security guards were also there, and in Argentina it is an open secret that the Security Guards will loot anything that is lootable. We tried to get in to rescue our computer and the work which had been left on it for the new issue, but the smoke made this impossible. The second attempt, behind the guards' backs, succeeded. Meanwhile, announcements were being made by sheer shouting; there was no megaphone.

Derek McNally contacted us. Could we get the last issue ready? We said we could. Then what about the venue for the posting of the Resolutions – essential IAU business, and the crux of the whole Assembly? We suggested that this had better be in the Plaza. By now we were standing forlornly on the sodden steps of the smoke-filled Centre, with what equipment we had managed to rescue. The IAU computer was still inside – so far as I know, it was never got out.

I talked to Derek, who asked me: 'Are you really going to be able to get the last issue out?' 'Yes.' 'Can you give the time and place where the delegates must meet to pass the Resolutions?' 'Yes. Plaza, 10 o'clock.' 'Correct.' 'I'll tell them that. By the way, where do they collect their newspapers?' 'Oh – the Plaza, at 10 o'clock.' It was the perfect example of circular reasoning.

We went back to the hotel, plugged the computer into the wall socket in John's bedroom with the lethal live wiring, and finished the last issue – minus the Resolutions, which of course had not been ready in time, and were now marooned in the abandoned, smoke-filled Centre. In the early hours of the following morning we dumped copies in the main hotels where the delegates were staying, and I must say that this worked; everyone turned up. The Resolutions had to be passed verbally.

The closing banquet then took place. There were only one-third the numbers required of chairs, knives and forks. *En route*, John's cab hit a pedestrian. As the main banquet was ending, a carvery was set up in the middle of the dance floor; this made dancing very interesting, though it didn't affect me, as dancing is not my forte.

The General Assembly ended with the closing ceremony, and we went to the airport to take the plane back to Los Angeles. The plane was nine hours late. Naturally, the bank at the airport did not cash travellers' cheques, and everything else was closed. Finally we got away, and in Los Angeles came back to civilization by boarding a London flight on Virgin airlines. As one of us said, we were never more glad to get into a Virgin…

I am glad to say that *Cruz del Sur* was well received, and it is true that many people said that it was the best IAU newspaper for many years. (Incidentally, all the spelling and phraseology were proper English; thus 'colour' instead of 'color'.) The research papers were excellent, and we had the greatest possible help from their authors. There was, in addition, one episode which deserves to go on record. This concerns the reported discovery of a new planet, moving not round the Sun, but round a pulsar several thousands of light years away.

A pulsar is produced when a very massive star – far more massive than our Sun – exhausts its nuclear 'fuel', and collapses. There is an implosion, followed by an explosion, and almost all the star is blown away into space, leaving only a tiny, super-dense core spinning round quickly and emitting pulsed radio waves, which we can pick up with radio telescopes such as the 250-foot 'dish' at Jodrell Bank in Cheshire. When I say 'super-dense', I really mean it. Fill an eggcup with material from a pulsar, and it will weigh more than a dozen ocean liners combined.

On 24 July, at Buenos Aires, Sir Francis Graham-Smith, the former Astronomer Royal, came into our editorial office and told me about the new planet, discovered from Jodrell by Professor Andrew Lyne and his team. I have to admit that I was instinctively sceptical. A pulsar seemed to be the last sort of object to be the centre of a planetary system, but who was I to quibble? Andrew Lyne is one of the world's leading radio astronomers, and my knowledge of that branch of science is, to put it mildly, sketchy. I did ask one question: 'What is the orbital period?' – that is to say, the time taken for the alleged planet to make one circuit of the pulsar. Answer: 'Six months.' Warning lights flashed in front of my eyes, and I passed a hasty note to John, who was standing next to me: 'This is a reflection of the Earth's orbit round the Sun.' But we collected the data, and ran the story on the front page of the next day's issue of *Cruz del Sur*.

Sadly, I proved to be right. The Jodrell Bank astronomers had made an error; they had omitted to take into account that the Earth's orbit is not a perfect circle, but an ellipse. It was an elementary mistake, but it made all the difference. The situation was cleared up at a meeting of the Royal Astronomical Society in London some weeks later. I was there when Andrew Lyne made his statement, and I have seldom admired anybody more. It would have been so easy to put the blame on some computer error, or an instrumental defect, but he did nothing of the kind; he said, quite simply, that he and his team had made a mistake. It was a classic example of scientific integrity, and in my view Andrew emerged with more credit than he would have done if the planet had really existed. (*En passant*, since then several more pulsar planets have been reported, though not from Jodrell Bank. I can only say that I remain sceptical.)

I will pass briefly over the next two General Assemblies, Holland 1994 (which I attended) and Japan 1997 (which I didn't). Then, in August 2000, we were back in England, this time in Manchester. I was again sounded about the Editorship, but at my suggestion John and I reversed rôles; he became Editor, with me as Assistant. Variety is the spice of life; moreover John is a far better organiser than I am, as well as holding all the right qualifications. Producing *Northern Lights* was less traumatic than *Cruz del Sur*, and all went well; the third member of our team – Chris Lintott, then in the midst of his degree at Cambridge – was invaluable. Between the three of us I know we did a good job, and the newspaper was truly professional as well as being entertaining.

Regretfully, I have to admit that the General Assembly at Manchester has been my last. I will not be at Sydney in 2003, or China in 2006. However, I am proud to have been the IAU's only amateur official, and certainly I will never forget those hectic three weeks at Buenos Aires, culminating in our successful attempt to salvage our computer from the burning San Martin Centre before the looting Security Guards had had time to move in!

17 Vote, Vote, Vote?

Political arguments have always been refreshingly absent from the IAU, at least in my experience, but of course the development of astronomical research, and in particular the development of space research, does depend upon politics and finance. If we are to fulfil our potential as the only thinking race in the Solar System we must rely upon world leaders, and I have to say that looking around the present world leaders inspires me with a feeling of no confidence at all.

I am certainly not a political animal, and only twice have I been involved in active campaigning. The first occasion was during the General Election in 1979, when I supported an Independent candidate. Colonel Edmund Iremonger, who stood for the Chichester constituency, held of course, by a Tory (Tony Nelson) who finally resigned a few years later and at once defected to the Labour Party. Edmund was an amazing character. He had had a highly distinguished war record, and held unusual medals from places such as the Middle East, about which he was always reticent. His manifesto was honest and straightforward: Put Britain First. The current Conservative leader was Edward Heath, who was obsessed with Europe, and naïvely believed that we could trust the French and the Germans, which did not please Edmund (or me). At the election Edmund did quite well, but, predictably not well enough to make a real dent in the Tory majority. When Mrs. Thatcher became Leader, it was clear that her manifesto was very close to Edmund's, and that was that. (I may add that the rules about alcohol in my home are the same as those laid down by Edmund, and are still referred to as the Iremonger Rules. They are quite simple: you help yourself.)

My second foray was much more recent – at the last General Election and again in Chichester, where Tony Nelson had snuggled down with his new friends in the Labour Party and had been succeeded by a man whose name I can never remember and whom I have never met. The United Kingdom Independence Party, UKIP, had come into being, and put up a good candidate, Douglas Denny. He did at least win enough votes to hold out some sort of hope for the future. Chichester is one of the safest Conservative seats in the country; someone said that if the Tories put up a chimpanzee as a candidate he would win – and someone else unkindly said that they had.

Mind you, I voted Tory for most of my career, and it was once even vaguely mooted that I might stand, but I said 'no' immediately, because (a) it would have meant giving up everything else, and (b), I always say exactly what I think. Around 1950 I did have a brief flirtation with the Liberals, who, believe it or not, were then a small but sensible party. I left when the young Liberals were invaded by people such as Peter Hain, now a member of the Labour Cabinet (!). It has to be said that the Liberals have had a curious history. I first remember them as being tacitly aligned with the Conservatives, and we had the National Liberals, who were indistinguishable from the Tories. Then came the Lib-Lab alliance, to keep the Tories out. This was followed by that comic-opera organization the SDP, the only lasting effect of which was to make the Liberals change their name to the Liberal Democrats. Now they are back with Labour. It is fair to say that if they thought they could win a few extra votes, they would happily join up with the BNP or the Socialist Workers' Party!

I have been accused of being ultra-Right wing, and in one

broadcast it was suggested that I stand politically rather to the Right of Genghis Khan. But this is not so. Far Righters are supposed to favour policies such a capital punishment and blood sports, and in that case I do not qualify. For example, I have already said what I feel about hunting animals with dogs, and I cannot go along with capital punishment. It would certainly reduce the murder rate, whatever the do-gooders say, but for me it is too risky, and mistakes can be made – for example Evans was certainly wrongly hanged and Hanratty possibly was, while the mentally retarded Derek Bentley did not mean to kill anybody. Also, I would not myself be prepared to execute anyone in cold blood, and it would therefore be moral cowardice for me to ask someone else to do it on my behalf. So I am in no sense a 'hanger and flogger'.

Do you remember the song by Flanders and Swann, 'The English, the English, the English are best?' I totally agree. Filling in the form at the 2002 census – which I did not in the least mind doing, because I have nothing to hide – I firmly crossed out 'British', and substituted *English*. After all, that is what I am, despite my Irish name. We are always being told that we must 'integrate'. I am sorry; I have not the slightest wish to integrate with anybody. I am proud of being English.

There are drawbacks to living in a country where there are only two effective political parties. It works excellently when both parties have good, honest leaders, but in our country it has not happened for a long time now (in my lifetime I don't think it has ever happened). New parties are always submerged by the power of financially backed propaganda, and this is a pity. It would be beneficial to give scope for new ideas.

I have a great deal of sympathy with the Monster Raving Loony Party, which differs from all the others in one vitally significant respect: its members *know* that they are loony. There are some splendid pieces of proposed legislation; for example, since Monday is the most unpopular day of the week, let us abolish it and have two Fridays instead. Since the regretted death of its founder, Screaming Lord Sutch, the Party has had co-leaders, one of whom is a cat, Mandu. She is of course a very intelligent cat, and when provided with a feline interpreter will no doubt be able to make useful contributions. My favourite question asked by the MRLP during a radio broadcast was addressed to a supporter of the Monopolies Commission, who was extolling the virtues of competition. He was taken aback when asked why, in this case, was there only one Monopolies Commission?

Mavericks do succeed occasionally, and not long ago a County Council – I forget which – discovered that a newly elected member was there solely to protect against the dreadful policy of putting fluoride in the drinking water. Mr. Charles Cockell, who fought a General Election seat on behalf of the Forward to Mars party, did not fare so well, and lost his deposit. His failure may have been due in part to the fact that his opponent was the then Prime Minister, John Major. Perhaps he will try again.

Are there any really honest MPs? The answer is 'yes', but, alas, they are an endangered species, and are in danger of extinction. I have known a good many politicians, some of them well, and I do appreciate that they find it difficult to remain sincere and at the same time obey directives from their leaders. If they rebel, trouble lies ahead. Enoch Powell was bitterly attacked because he said what he thought; looking at the situation as it is today, can anyone please tell me where he went wrong? Whether you approve or not, nobody can honestly deny that most of what he predicted has actually come to pass. I liked Harold Wilson, who was not without a sense of humour; at a reception in No. 10, which I was at, he filled his pipe from my tobacco pouch and laughed when he realized that he was about to smoke best Rhodesian. I admire Tony Benn; my political views differ from his by about 180 degrees, but nobody can doubt his decency or his sincerity. And I admire Margaret Thatcher, who was the right person in the right place at the right time, just as Winston Churchill was in 1940. During the

Falklands War, our people were under attack; we hit back, and drove the Argentines out. If we had any other peacetime Prime Minister, General Galtieri would have got away with it. Edward Heath would have dithered; Harold Wilson would have let events take their course; Alec Douglas-Home wouldn't have known that it was happening, and our present Government would have been afraid to take any action without permission from their masters in Brussels. So it is just as well that Mrs. Thatcher was in charge.

Yet even her Government did not have an unblemished record, and was guilty of the betrayal of Rhodesia. Whatever you think about 'white rule', the Rhodesia of the 1970s and 1980s was peaceful and prosperous. Left on their own, the Rhodesians would steadily have broken down the barriers between the races. Unfortunately nothing would satisfy our Labour Government except the overthrow of the régime, and the Tories to their discredit, kept up the pressure. The country was handed over to Mugabe, a thug of the worst type. Today our present Government is plotting to betray Gibraltar in the same way, but not for a moment do I believe that the Gibraltarians will surrender. After all, they didn't surrender in 1940.

If I had to select one country which had an excellent political system I would go for Liechtenstein – all 79 square miles of it. I had the whole story from one of the close relatives of the Ruling Prince (I met him in a shop in Vaduz, the capital; I had gone in to buy some camera film). There are two political parties, and elections are held every four years, but as the aims and objectives of the two parties are identical nobody bothers to vote and the Ruling Prince goes on ruling, financing the administration by postage stamps, a false teeth factory and the manufacture of cuckoo clocks. There is a police force – he's a very pleasant chap – and a prison, which I gather is used as a supplementary hotel during the tourist season. I put one question to my charming acquaintance. 'Look, you have been an independent state ever since the thirteenth century. How have you managed it?' He gave me a knowing smile. 'It is easy, my friend,' he said. 'We are no bloody good to anybody!'

Perhaps there is a moral there ...

America? Well, I have spent a good deal of time there, and in my Moon-mapping days I was a very frequent visitor, but I have to admit that when I am in the United States I have the feeling of being cut off, because nothing which happens in the rest of the world seems to be regarded as of the slightest importance (if you doubt me, try to find out the cricket scores when you are staying in New York). Occasionally the outside events percolate in, as during the Gulf War, when a concerted attack was made on Iraq, following the invasion of Kuwait. Much was then said about the righteousness of it, and the need to protect small nations. Yet, I fear that if Kuwait had exported coconuts rather than oil, nobody would have cared tuppence about it.

The only American politician whom I have known faintly well is Ronald Reagan, for whom I have tremendous regard. Those who regarded him as a clown were making a very big mistake; he was a very clever man, and he will go down in history as one of the great US Presidents. I love his sense of humour, and I remember hearing him speak at a meeting when he was campaigning for George Bush (the first George Bush). 'When I was running for President, some very unkind things were said about me. I was even accused of being an actor!' The audience fell around. He also told me that every morning he looked at the Obituary notices in the paper. If he wasn't there, he went on with the day's work.

It is tragic that he was struck down with Alzheimer's disease; when it was diagnosed, he accepted the situation with courage and resolution. The world was a safer place when Ronald Reagan was in the White House.

I once went to America with David Hight, of the publishing firm of Mitchell Beazley, they had just put out a book of mine (*Atlas of the Universe*) and we were doing some publicity for the US edition. We flew to New York, and our next port of call was

Washington. For some reason or other we decided to go by train, which is not something which one usually does in the United States. We settled ourselves at comfortable seats at the bar, and ordered some drinks (for purely medicinal purposes, of course). It was all very pleasant, and I think we dozed off. We woke to hear an announcement: 'Washington'. Hurriedly we collected our bags, and alighted; the train moved away. We walked down a tunnel, and emerged into the countryside, with cows saying, 'Moo'. 'I don't' think this is quite right,' said David.

It wasn't. We had disembarked at the state border, which is a long way from the city.

There was no train for several hours, but we managed to contact our hotel to say that we would be late; the hotel was the Watergate, which at that time was quite unknown. The hotel manager was very helpful, but explained that as we had not confirmed our booking the rooms were no longer available. 'But can't you fit us in anywhere?' A pause. 'As you are visitors to our country, I will accommodate you. You can have the Presidential suite, which isn't being used tonight. Glad to help.'

So that is what happened – we stayed in the suite which later became so notorious. As we still had time to spare, we decided to look round Washington, but it was now dark, and at such a time it is not wise to go around on foot. We inquired at the Rockefeller Center, right opposite the hotel, and asked if any entertainment could be found there. We were assured that there was. 'Yes, a recital by two Jewish tenors singing songs in Yiddish. Do come – there's plenty of room!'

Alas, I'm afraid that we did not take up the invitation, though I have no doubt that the songs were of the highest calibre.

I have every wish to keep in friendly touch with Europe, but I definitely do not want to be part of anything like the United States of Europe, which would be dominated by the Germans, the French and the Italians – the very people whom we fought, and beat, sixty years ago. I am not taken in by 'Eurobabble', and I prefer to remain independent, as we have been for the past thousand years. I may be accused of being a dinosaur, but I would remind you that dinosaurs ruled the world for a very long time. Note also that Norway, which has had the common sense to keep out of the European Union, is doing very well indeed.

How to vote? Not long ago I was peripherally involved in a programme about politics, during which we sent out a form asking viewers the name of the man or woman who they would really like to see running Britain. Hundreds of votes were cast. Tony Blair did quite well, and so did Margaret Thatcher, David Beckham and various pop idols – but the winner, by a mile, was Guy Fawkes.

18 The Battle of Greenwich

There are politics in science, as in everything else, and there are also personal quarrels, in which I am very careful not to get involved. As an amateur, without a conventional degree (thanks to Herr Hitler), but one who has taken part in a good many professional programmes, I could easily have been a target. In fact this has not happened, and I have never been attacked by any professional astronomer of real standing. This may well be because I have had the sense not to delve too deeply into matters about which I am not qualified to make personal contributions. For example, I am not a cosmologist, and I am not a good mathematician; I would never claim to be. All I can do is to try to interpret the results obtained by others, so that they can be made accessible to non-scientists. Only with regard to the Moon and planets do I feel that I can put forward my own views, and even here I have made mistakes (in particular I was convinced that the Moon's craters were volcanic, and it was a long time before I had to admit that they were formed by meteoritic bombardment.)

However, I have occasionally run foul of second-raters, and one episode was rather amusing. I had written a book called *Suns, Myths and Men*, which came out in 1956 and was concerned with the history of astronomy, including some old ideas – such as the Indian theory that the flat Earth was carried on the back of four elephants, which in turn stood on the back of a large turtle (I have always felt sorry for the turtle, which must have been in serious danger of being turned into turtle soup). The book was 'reviewed' by one H. C. King, who later worked at the London Planetarium for a few years after I had turned down the Directorship. He certainly did not like me; I was never sure what he was getting at, but he was entitled to his opinion. This was the very book that television producer Paul Johnstone read, at the same time, and which led him on to invite me to put on *The Sky at Night.* No comment!

Even though I am astronomically apolitical, there was one occasion when I became heavily involved in what was essentially a political episode. This was the battle to save the Royal Greenwich Observatory.

The RGO was originally founded in 1675, which was rather before my time. Oddly enough, its initial function was not purely astronomical, but navigational. Britain has always been able to navigate; this is easy enough today, with all our modern devices, but it was very far from easy in the seventeenth century. Once a ship was out of sight of land it was liable to lose its way completely, sometimes with tragic results.

To fix one's position on the surface of the Earth you need to know your latitude and your longitude. Finding latitude presents no difficulty at all; you simply measure the altitude of the Pole Star above the horizon, make a slight correction to allow for the fact that the Pole Star is not exactly at the celestial pole, and the problem is solved; the altitude of the celestial pole in degrees, minutes and seconds of arc is the same as your own latitude on the Earth. This of course applies to the northern hemisphere. There is no bright South Pole star, but sailors in the Middle Ages could always work out latitude with reasonable accuracy. The real problem was in finding longitude.

A ship's longitude is the difference between the meridian it happens to be on, and a standard meridian, such as that of Greenwich. Local noon is easily found, since this is the moment when the Sun is at its greatest height above the horizon. A reliable clock will give Greenwich time at this moment, and so the longitude of the ship can be calculated. Unfortunately sailors of the seventeenth century had no reliable clocks, and it was not for many years that John Harrison, son of a Yorkshire carpenter, developed a chronometer which was adequate for use at sea. Before that, the only real way to find longitude was to use the Moon as a sort of clock-hand. It moves against the background of stars, and if you know exactly where it is at any moment you also know the Greenwich time. This does not involve using any sort of mechanical timepiece, but it does mean using a highly accurate star catalogue, and this was where the trouble began.

The best available catalogue had been drawn up between 1576 and 1596 by Tycho Brahe, a curious Danish astronomer whose other claim to fame was that after having part of his nose sliced off in a student duel, he made himself a new one out of gold, silver and wax. He had no telescope; instruments of this kind did not come along until twenty years later, and so Tycho had to make all his observations with the naked eye. He did remarkably well, but even so his catalogue was not precise enough to satisfy the navigators.

Telescopes made all the difference. By the mid-seventeenth century they had become effective, and a new star catalogue based on telescopic measurements became possible. Enter the Royal Society, founded in 1660, and enter King Charles II, who had a lively interest in science and was warmly supportive of a plan to establish a major observatory specifically for navigational requirements. He even financed it, paying for it by selling 'old and decayed gunpowder' to the French. (I have always admired Charles II; he was an effective ruler, his morals were strictly comparable with those of present-day Royals, and his methods of dealing with the French were eminently sound; we of 2003 would be wise to follow his example.)

The RGO came into being, and the first Astronomer Royal, the Rev. John Flamsteed, did produce the required catalogue, though he had to buy his own telescopes and his sole assistant, a 'silly, surly labourer' named Cuthbert, was much more at home in the local tavern than in the Observatory. Thereafter the RGO (Royal Greenwich Observatory) was the centre of British astronomy, and during Victorian times it acquired telescopes which were powerful by the standards of the day. For many years the Director was George Biddell Airy, a curious character who was obsessed with 'order and method'; he once spent a whole day in the Observatory cellars labelling empty boxes 'Empty', and he insisted that his observers should remain on duty throughout the night even when rain was falling. (It is said that his ghost still wanders around among the domes after nightfall. I have never seen it myself, but I know people who say they have!)

Airy was mainly responsible for making Greenwich Observatory the world centre for longitude, so that the prime meridian (longitude 0°) passes through it and divides the globe into two hemispheres, you can straddle it with one foot in the eastern hemisphere and the other in the western. This was settled in 1885, when international agreement was much easier to obtain than it is now. The main opposition came from (surprise, surprise!) the French, who wanted the prime meridian to pass through Paris. It was years before they gave in and adopted Greenwich Mean Time.

The RGO became the centre for pure astronomy as well as navigation, but gradually the Royal Park became unsuitable as a site, because London was becoming not only larger (which did not much matter), but also brighter (which mattered very much indeed). Light pollution was an increasing problem, and eventually it was decided to move the main equipment down to the darker skies of Herstmonceux in Sussex, leaving the old Royal Observatory as a museum. Herstmonceux Castle became the administrative centre, and the wheels were set in motion by the Astronomer Royal at the time, Sir Harold Spencer Jones. The Second World War delayed matters, but by 1956 the shift was complete; Spencer Jones retired, and was succeeded by Sir Richard van der Riet Woolley. I knew both well; in fact I broadcast with Sir Harold a day or two before his sudden death, and I was staggered when I received a phone call asking me to go straight to Broadcasting House to give an appreciation of him. Dick Woolley was a forthright character; a world leader in astronomy, and, incidentally, a very good cricketer and tennis player. (I used to enjoy bowling at him in the nets; the last time I faced him on the tennis court he won 6-1, 6-2).

He retired in 1970, and the Science and Engineering Research Council, which controlled the finances, made a classic error. Up to that time the Astronomer Royal had also been Director of the Royal Greenwich Observatory, but now the two were split; Dr. Margaret Burbidge became Director of the RGO, while the title of Astronomer Royal went to Sir Martin Ryle, who was a radio astronomer and who was based in Cambridge. This did not look as if it would work – and it didn't. Margaret Burbidge resigned after a brief tenure of office, but the post of Astronomer Royal never returned to Greenwich. In my view, this was really where the rot started.

However, for a while things went on more or less normally. Two things happened, only one of which made sense; this was the decision to move the Observatory's largest telescope, the 98-inch Isaac Newton reflector, away from Herstmonceux to the clear skies of La Palma in the Canary Islands, where an observatory had been established on the summit of an extinct volcano, the Roque de los Muchacho. (At least, I hope it is extinct; there seems to be a certain amount of doubt about it, and certainly there is active vulcanism elsewhere on the island.) The INT (Isaac Newton Telescope) was to be given a new, more modern, mirror; the mounting had to be drastically altered, because La Palma is well south of Greenwich, and of course, there would have to be a completely new dome. It would have been wiser to leave the original INT where it was, and build an entirely new telescope for the Canaries, but this would have cost slightly more, and the accountants – those banes of the modern world – were on the alert. The upshot was that after the move Herstmonceux was left with a 98-inch mirror, most of the telescope mounting, and a dome.

This gave me an idea, and I went to see Professor Alec Boksenberg, who had become Director of the RGO though without the title of Astronomer Royal. (As a designer of scientific instruments he is in a class of his own; he was largely responsible for the development of the CCD or Charge-Coupled Device, which is now invaluable – why he has not been given a knighthood passeth all comprehension.) My idea was this. Why not re-erect the telescope at Herstmonceux, making new parts where necessary, and make use of it?

There would be several advantages. First, the telescope would still be the largest in Britain, giving Herstmonceux distinct prestige value. Secondly, it could still be used for research. Thirdly, it would enable up-and-coming astronomers to gain experience in operating large telescopes. I well recalled how David Allen had told me that when he was first given observing time on the Palomar 200-inch reflector, in California, he was able to get down to work straight away, because he had cut his teeth on the INT and knew exactly what to do.

As I had expected, Alec was receptive, but as usual he had to consider the accountants. I asked whether I could investigate: he said 'Yes'. I cast around, and found a Scottish engineering firm ready to re-erect the telescope on a revised mounting, and pay the full cost, provided that the telescope could be named after them. This sounded fine. The Isaac Newton Telescope was now in La Palma, if the Herstmonceux 98-inch became the Buggins Liver Paté Telescope, who would object? At Herstmonceux, a committee was set up to oversee the proceedings, and work was just about to start when the real blow fell. The SERC (Science and Engineering Research Council) now intended to close Herstmonceux altogether, and re-locate the RGO in Cambridge in what was tantamount to an office block.

In fact there were several suggested venues; others were

Manchester and Edinburgh, but Cambridge was clearly the front-runner so far as the SERC was concerned. Needless to say, the astronomers at Herstmonceux did not want to re-locate anywhere, because it was patently obvious that they would lose all prestige as well as their facilities (uprooting the telescopes and shifting them to the dank climate of Cambridge was a little too fatuous even for the SERC). Clearly, a battle was imminent, and virtually all the astronomers were up in arms, but they were handicapped in as much as they were Government employees, and free speech was not encouraged. Indeed, some of the more outspoken astronomers were officially threatened with disciplinary action. Fortunately I heard about this at an early stage, and wrote a letter to the SERC authorities: 'Take this one step further, and I call a Press conference and blow the whole thing up in your faces.' I wasn't bluffing. Evidently, this was realized, and there was no more talk of disciplinary action.

You may well ask why I, as an amateur, was playing a rôle in all this. The answer is that I was asked to do so, and although I was a full member of the International Astronomical Union, I was not subject to any disciplinary action, to that I was able to take the lead. Accordingly, and with Alec's full approval, I booked a room at the Science Society's Lecture Theatre (where both the RAS and the BAA held their meetings) and sent out a circular, inviting all leading astronomers to take part – plus, of course, the hierarchy of the SERC. I felt that it would be inappropriate for an amateur to take the chair, and so I telephoned Professor Patrick Wayman, Director of the Dunsink Observatory in Ireland, and asked him to come over. He agreed. As he was one of the world's leading astronomers, this was clearly a wise move.

Response was interesting. The astronomers flocked in, but the SERC was reticent, and the Chairman, Professor Mitchell, suddenly found that he had an urgent appointment elsewhere. Eventually one SERC representative did turn up, looking decidedly embarrassed.

Chris's photo of the total solar eclipse in the Caribbean.

Chris preparing for the eclipse on the deck of the good ship Stella Solaris.

With Chris in Antarctica, 1999.

With astronauts, 2000.

Investiture at Buckingham Palace, with Adam (left) and Chris (right).

With Buzz Aldrin at the BAFTA Awards, 2001.

With young helpers at the South Downs Planetarium, 2001.

The BAA Council, Burlington House, 2001.

18 Jeannie in command! My study, Selsey 2001. She generally sleeps on the tray indicated . . .

With Brian May at Selsey, 2002.

With Jeannie, 2002, at Farthings; photo by Brian May.

With John Watson; testing small telescopes at Farthings.

My desk at Farthings, the Woodstock is much in evidence.

The meeting was held on 6 June 1986. I gave a brief introduction and then handed over to Patrick Wayman, who was – predictably – an ideal chairman for the occasion. Speaker followed speaker; one stalwart was Garry Hunt, the planetary meteorologist, who was a major figure in the space research programme, and whose views carried a great deal of weight. Significantly, the Director and Secretary of the Cambridge University observatory spoke against any move; they did not want the RGO on their doorstep.

Another factor was the effect on the Astronomy Department of the University of Sussex, one of the strongest in the country. Most of their PhD students had made full use of Herstmonceux, and this was also true of students from other universities. Remember, the Castle was unique.

Finally, a vote was taken. The result was that not counting the SERC representative, who kept prudently silent, not a single vote was cast in favour of the move. It was absolutely decisive, and I think that we all went away believing that we had done enough to get the decision reversed. In my case I went straight from the meeting to King Edward VII Hospital in Midhurst, where I had to have a very nasty thyroid operation; I had actually managed to postpone it, against medical advice, until after the meeting had been held (which, as a private patient, I could do). For the next few weeks I was totally out of action. I still have the large get-well card sent to me and signed by most of the astronomers who had taken part in the resistance.

Our hopes were dashed. The Ministry took not the slightest notice of the astronomers, and the move went ahead; all the portable equipment was taken away, the telescopes were left to rot, and the Castle was deserted. It was a sad time.

I went there at a later stage, when most of the astronomers had left – many to retire, others to leave astronomy altogether (not very many expected to go to Cambridge). I arrived after dusk, because I was anxious to use one of the telescopes for a special observation, and had been given full permission to do so. It was rather eerie; there was a feeling of gloom, and it was hard to believe that only a few months earlier it had been Britain's flagship observatory.

At the very end, a farewell party was held in the Castle, and for an hour or two it was hard to credit that this disaster was really happening. As I remember saying, it would have been a splendid party on any other occasion – but it was the last.

Subsequently, the Castle was put up for sale, and there were some curious manoeuvres about which I know very little. At one stage everything was taken over by a man whose name I forget; apparently he planned to convert the Castle into a luxury hotel, complete with swimming pool and children's playground. I instinctively classed him as an educated barrow boy, and apparently I was right, because he soon went bankrupt and vanished from the scene. The situation was saved when the Queen's University at Ontario, in Canada, bought the Castle for use as a study centre, and leased the telescopes to a newly-formed Science Centre. Two of the telescopes are now back in use, and at least there is still astronomy at Herstmonceux.

Meanwhile, the transferred RGO was re-established at Cambridge, minus its telescopes, its library, its priceless plate collection, and much else. Rather to my surprise, I was invited to the official opening, and I have a photograph showing me together with Mr. Jackson, from the Ministry of Science. We are standing in front of the plaque, shaking hands, smiling pleasantly at each other, and each thinking: 'Drop Dead!'.

My fear – shared by many others – was that the RGO would be quietly absorbed into the Cambridge Institute of Astronomy, and would cease to be a truly independent body. At that time Mrs. Thatcher was Prime Minister, and I wrote to her. I quote her reply, dated 7 June 1986:

'All the present RGO activities will move to Cambridge apart from the Equatorial Group of telescopes and the Exhibition

Centre. Thus the Observatory will retain its independence and integrity, and there can be no question of the loss of the RGO as a national observatory.'

That seemed clear enough. Yet little more than a decade later, in 1997, the Government announced that the RGO was to be closed completely. Not even the name would be retained, and moreover the Observatory at Edinburgh would no longer be Royal.

Even I was taken aback, and initially I simply could not believe it. Alas, it was true enough. Massive opposition was mustered, and I was again asked to play a major role, simply because of my amateur status. There was support from both Houses, and several meetings were held, though the officials of PPARC (successor to the SERC) were decidedly coy about attending. The Civil Service was too strong, and on 30 April 1998 the Royal Greenwich Observatory ceased to exist. It was not modified, mothballed or merged; it was destroyed.

In April 1999 I published a letter in the *Observatory* magazine, an official organ of the Royal Astronomical Society, and in this I set out what I thought about the closure of the RGO. I quote:

> We have lost:
> The RGO as a major administrative astronomical headquarters. Its closure has left a vacuum. The loose organization which also involved what used to be the Royal Observatory Edinburgh cannot effectively fill this rôle.
>
> The expertise of the RGO team. It had been damaged by the move from Sussex, but it was still second to none. It has been said that none of the members of the astronomical team have been made redundant. (In fact 66 of the 107 RGO staff *were* made redundant.) Nevertheless, they are dispersed, and many of them have moved away from astronomy. (*En Passant*, the telescopes at Herstmonceux were neither the world's largest nor the world's best, but they were of immense value for testing new equipment. I once

asked a leading official of the SERC about this. His reply: 'Take the equipment to La Palma and test it there.' Clearly he had never heard about the pressures on telescope time.)

Papers. The RGO published around a hundred per year, and it was officially recognized that some were among the most cited papers internationally.

Instrument development. The RGO's success with, for example, the William Herschel Telescope has led to its very close involvement with Gemini and other major projects. This can hardly continue now.

Education. The RGO has always been at the forefront of the astronomical education, and though this was truncated when Herstmonceux was lost, it did continue at Cambridge.

The Library. This may well survive, but not as a separate entity, but merged with that of the Institute of Astronomy. It must be made available to all researchers of astronomy, with duplicate copies being shipped to the La Palma Observatory.

The priceless plate collection. I am told that these plates are now boxed up and stored (at great cost) by a professional firm in London, so that nobody can use them.

Herstmonceux as a conference centre. It was in use almost all the time, both for technical and specialised gatherings and for education. The future of the Herstmonceux astronomical conference series and the student vacation course, which both transferred to Cambridge, are uncertain.

Heritage. The Royal Observatory was founded by King Charles II in 1675, and has remained the world's most famous observatory ever since. Overseas colleagues simply cannot credit that it has been dismantled. Moreover, the RGO staff wanted to preserve the name and maintain a nucleus, so that in the fullness of time the RGO itself could be brought back to life. The Government made very sure that this did not happen, and even the purpose-built offices are being disposed of at breakneck speed.

I understand that the initial cost of closing the RGO is around £8,000,000. In the end, of course, no money will be saved, because what was being spent yearly on the RGO will have to be used in compensatory activity elsewhere – unless British astronomy is to end completely. The loss of status of the Royal Observatory Edinburgh has not yet become widely known, and this is a factor to be taken into account.

What can be saved from the wreckage? Well, the name may be preserved if it goes to the museum at Greenwich, but this may have to be modified – and it is worth noting that the money being spent on the huge, useless Millennium Dome at Greenwich would have financed the RGO well into the 22nd century. All we can really do now is to build upon all that the RGO achieved during its 300 years of existence.

I do not for a moment doubt that those concerned believe that they are doing the right thing for British and world astronomy – but neither do I doubt that when the story of the 20th Century comes to be written, the closure of the RGO will go down as the worst of all acts of scientific vandalism.

I did make one final, if minor contribution to the story of the RGO. After the closure, I suggested to the last Director, Professor Jasper Wall, that it would be a good idea to found an RGO Society, to hold meetings, to make sure that the name was preserved, and to re-establish the Observatory when possible. Jasper was enthusiastic; the Society was formed, and I was honoured at being invited to become a member. Fittingly, the first big meeting was held at Herstmonceux in 2000; it was well attended, and despite the sense of nostalgia it was a cheerful occasion.

Will the RGO ever be revived? It is difficult to say. My acid comment was: '"RGO-RIP": founded by Charles II in 1675, destroyed by Sir Humphrey Appleby in 1998'. But I hope that things will change, and that we have not heard the last of the great national observatory which led the way for so long.

19 Crazy Consignia

It was, I suppose, during the 1980s that I began to realize that I was facing a new situation. By that time I had been appearing on television, at least twice a month, for more than twenty years, and the effect was bound to be cumulative; people did recognize me, and I became used to carrying a stack of personal cards to give to youngsters who wanted autographs. I well remember that once when I was in Greece, and was walking round the Acropolis (something I gather you can't do today). I signed more than a dozen cards for teenagers and pre-teenagers.

Absurdly, anyone who appears regularly on television is regarded as both a celebrity and a pundit, I am neither. *The Sky at Night* has lasted for so long not because of any skill on my part, but because of the fascination of the subject; there can be nobody who does not take at least a passing interest in the sky above. There are many presenters who could fill the rôle better than I can, but it so happened that thanks to Paul Johnstone I was first in the field, and there I have remained, I know that without meaning to be obstructive I block other would-be presenters, some of whom are anxious (even over-anxious!) to replace me, but this is a problem to which I can see no easy answer.

Neither am I a pundit. I am an astronomer, and the fact that I know a little about the Moon does not qualify me to give my views about, say, the economy of Swaziland or the parlous state of the railways in Novaya Zemlya. However, it is only too easy to give a wrong impression. Years ago I was taking part in the BBC television programme *Panorama*, talking about something connected with astronomy, when the producer came up to me. 'We've booked an expert to talk about feeding stuffs for animals – we're due on in five minutes, and he hasn't turned up. Can you do it on autocue?' This was no problem, and I suppose I must have sounded convincing, because I had a stack of letters asking for advice about animal food – a subject about which my ignorance is complete.

Much more recently there has been *Gamesmaster*, a children's television programme about computer games. I acted as the presiding genius, and it was my role to guide the contestants, explaining how they had to dodge the hideous traps set for them in the shape of demons, dragons and assorted ghouls. In various episodes I was dressed as a sun-god, a sea deity and an Underworld spirit.... I dutifully read out what appeared on the autocue. What it was all about I had not the slightest idea, but for a long time afterwards I was besieged by young enthusiasts who wanted to know how they could escape from Level 6, etc. Rather lamely I explained that this was secret, classified information, and that they would have to work it out for themselves. (I became involved because the TV company said that if I would do it for them, they would give a substantial donation to the Cystic Fibrosis Research Trust. C. F. is a particularly nasty disease which affects children; mainly, though not entirely, boys. Obviously I couldn't turn them down, and anyway it was quite pleasant and harmless.)

I do have a great many letters, and the fact that the Royal Mail spent millions of pounds on changing its name to Consignia and back again has made no difference. So this may be the moment to say something about letters, phone calls and unexpected callers.

Some letters are decidedly odd. For example, I was intrigued by a Scottish correspondent who told me that his hobby was collecting the registration numbers of bassoons. I sent him one, and he wrote back: 'That's no good – it's a xylophone.' It was: So he really knew his subject!

There was also the writer with the grand old English name of Smith (I am sure this was genuine), who sent me a letter which seemed at first to be perfectly rational. He asked a couple of questions on astronomical matters, and then went on to inquire about the best way of obtaining a job in a scientific institute. The sting came on Page Three: 'The situation is complicated because I am the son of God, and should not therefore be unemployed.' I referred him to his local Vicar.

This sort of thing is harmless enough, but inevitably there are obscene letters too; I imagine that these are received by anyone who is at all in the public eye. In 99.9 per cent of cases the letters are anonymous, though I have had a few from fox-hunting supporters who did give their names (not their addresses). In my view, the only course is to adopt a policy of masterly inactivity. If you send even one reply, assuming that you know the address, you open the floodgates.

Many of my letters are from youngsters who want astronomical advice. These are always answered as soon as possible, and in full. In fact I never fail to answer a normal letter; my 1908 Woodstock works overtime, though if I am away for any reason, or if I am ill, mail stacks up and takes some time to clear. To speed things up I do have a number of 'standard' letters, and it may be worth giving some of them here.

First, the 'School Project Letter'. I have a great many inquiries, because astronomy is a favourite project, and few State school teachers know the difference between astronomy and astrology. So I reply as follows:

> Many thanks for your letter. Of course, I will help if I can. Your first step must be to do some reading, and make yourself familiar with the general background. Your School library may have books, if not, your local Public Library certainly will. Go

through a few of these, and see which parts interest you most. Then let me know how you get on, and I will do all I can.

Next, the 'Telescope Letter', sent to people who are anxious to fit themselves up with some sort of equipment.

> Many thanks for your letter.
>
> A great deal can be done with the naked eye alone. If you obtain an outline star map, and go out on clear nights, you will soon learn your way around; it does not take long. Binoculars are a great help. I do not recommend buying a very small telescope for a few pounds or a few tens of pounds; you would be better off with binoculars. I suggest contacting your local Astronomical Society [add the address]. If they cannot help, let me know.

Professional astronomy? Again, inquiries are common, and I have to give an honest reply, with my 'Careers Letter':

> Many thanks for your letter.
>
> Astronomy is open to everyone, and amateurs can – and do – play a useful rôle; no specific qualifications are needed. Professional astronomy is a different matter, and here a degree is essential. It may be in pure astronomy, but many enthusiasts prefer to take a first degree in physics; after all, modern astronomy is largely astrophysics. Either way, the first essential is to obtain your GCSEs and A Levels (including maths, physics and English). You can start to think about applying for a degree course. It will give you a fascinating career, and I wish you all success.

New cosmological theories come in at the rate of at least two per week. Some are bizarre, and I recall one theorist who maintained that the universe was shaped like a large cup; quaintly he added,

'This holds water, in spite of all criticism.' I think that I now have a satisfactory 'Theory Letter':

> Many thanks for sending me an account of your theory.
> I would like to help, but unfortunately it is impossible to make any useful comment without seeing the full, rigorous mathematical analysis. There is simply no other way to test whether a theory is valid or not, and there is no short cut. I am not a relativistic mathematician, but I know plenty of people who are, and if you send me the calculations I will have them properly refereed.
> I wish you all success in your research.

As the theorists never know any mathematics, this is quite safe, and in 99 per cent of cases nothing more is heard. The occasional philosopher will persist, and the procedure then is to send him a reply which includes a couple of mathematical formulae. This always works. Interestingly, almost all theorists are male; very few ladies are involved.

After a long period of experimentation, I have come up with what seems to be a valid 'Crank Letter':

> Many thanks for your letter.
> I hope you will forgive me, but pressure of time makes it necessary for me to declare a closed season on all correspondence pertaining to:
> *Astrology*
> *UFOs*
> *Creation theory*
> *The face on Mars*
> *Crop circles*
> *'Men never went to the Moon'*
> *Conspiracy theories*
> *Alien visitations from space*

I do hope you will understand.

With all good wishes,

Patrick Moore.

Away from astronomy, there are a few other constant inquiries and comments. Blood sports enthusiasts are with us, and I have my 'Hunting Letter':

Many thanks for your letter. I appreciate the fact that unlike most correspondents with similar views, you had the courage to give an address.

We will get rid of hunting this time; it is supported only by a well-organised and vocal minority. The trouble is that we cannot have a rational argument with these folk; cruelty and stupidity go together.

All good wishes.

Either there will be an obscene reply, or else none at all.

'Celebrity cookbooks' are all the rage, and people such as myself are asked to submit their favourite recipes. In 2002 I had a grand total of *174* similar requests. A standard 'Recipe Letter' became essential:

SELSEY RAREBIT

Take a slice of bread (white, in my case; I hate brown bread, which tastes like cardboard). Toast one side.

Butter the second side. On it, put chopped or sliced cheese, adding seasoning to taste (salt, pepper, mustard; Worcester sauce goes rather well). Toast the second side until the cheese is golden and bubbling.

Eat.

Two Selsey Rarebits, clapped together face to face, make one Martian Pancake.

I hope you enjoy it.

There is a story associated with this one. I was asked to take part in a television programme called *The Reluctant Cook*, in which the professional chef was to act as my teacher. He and the team came to Farthings, and asked what I would like to cook. I said 'Fish', and we went ahead. I may say that my current cat, Bonnie, had a splendid time, because all the fish used during rehearsals was jettisoned – and Bonnie was very fond of fish (unlike my present cat, my beloved Jeannie, who can't abide the stuff). At one stage I had to peel an onion, and the chef was very scornful about my method of doing so. He went on to demonstrate the correct procedure – and cut his finger.

The programme itself went out without a hitch, but viewers jumped to the conclusion that I was a really good cook. In fact I am not. If pressed I can put on a decent meal; I pride myself on my seafood salad, and I can make a good curry, but that is about all. I am always told that I make an African curry rather than an Indian one, because I believe in currants, raisins, sultanas and, above all, plenty of bananas.

There are folk who write for the sake of writing, and want to start a correspondence which will go on forever and achieve nothing. For them I have the 'Nebulous Letter':

Many thanks for your letter.

I agree! There is so much that we do *not* know, and although we have made great progress in recent years we are still doing no more than scratching the surface. Research is laborious and time-consuming; if only each day had 48 hours instead of only 24! But at least the future will be exciting. All best wishes.

Finally – the 'Star Names Letter'. Various bogus agencies claim to be able to allot names to stars – and they tend to target bereaved people. So I have a statement ready:

> Various agencies claim to be empowered to name stars, of course on payment of a sum of money. 'Send us £50, and you will have your own star.' DO NOT TOUCH THESE SCHEMES. Stars have not been named for centuries; the agencies are completely unnoficial. Have nothing to do with them.'

Let us turn next to phone calls. The telephone is an essential part of life, but it can be a mixed blessing. An answering machine is a help, but does not solve all the problems.

Advertisers can be irritating, but I have found a way of coping with them. If the phone rings, and the caller begins 'We are in your area, and if you need double glazing…' I cut in and say 'Do tell me about it.' I then simply leave the phone off the hook, and let them talk; they can chunter on for as long as they like, and of course they are paying for the call.

Obscene calls are unpleasant, but there is no real way to stop them, and the only course is to ignore them. Most of them are anonymous. I did once track a caller's number by dialling 1471, and went so far as to ring the local police station, but of course it was unmanned (all the policemen were out motorist-hunting), and I was far too busy to follow it up.

I am vulnerable to this sort of thing because my main number is not ex-directory, so that anyone can look it up in the phone book or, for that matter, in *Who's Who*. The reason for this is that many people, including youngsters, want to contact me when anything unusual happens in the sky, and I do not want to remain aloof. I had a case of this recently, in May 2002, when all the five naked-eye planets (Mercury, Venus, Mars, Jupiter and Saturn) were lined up in the western sky after sunset – something which will not happen again until September 2040. A 'planetary massing' of this sort is purely a line of sight effect, and means nothing, but it is spectacular, and caused great interest. I lost count of the number of phone calls I had about it, and I must have made dozens of broadcasts on local radio stations.

Unwelcome callers can be a nuisance. Of course most of them are religious fanatics (Jehovah's Witnesses and so on), and I have a few set gambits:

> No, I haven't been reading the Bible lately. I haven't much time for science fiction at the moment. Good-bye.
>
> Sorry – I'm a Druid, and I'm a busy Druid.
>
> I am on the Dark side – my number is 666. (This puts them to flight immediately. They can't get through the gate fast enough).
>
> I am not personally interested, but I know someone who would be very keen to discuss it with you. (Then give the name of someone with whom you have a score to settle.
>
> Once, when in East Grinstead, I put one caller on to the local curate, who had run me out in the match that Saturday when I had made a few runs and was, for once, feeling fairly confident. Subsequently he rang me up, and said, menacingly. 'Next time you send me any crackpots I will come round and *do* you.').
>
> To be used only as a last resort: 'I have only two words to say to you – and the second is 'off'.

I have never had any real trouble with trespassers, if only because I am very happy to show my observatory to anyone who wants to see it. But I recall the case of a friend of mine, Frank Hyde, who lived in a Martello tower on the East Coast, and was plagued by uninvited visitors. He put up a sign: *Radiation*, after which the trespassers kept away. The police then told him that his notice was illegal. Frank had the answer: 'This is the name of my house – nothing unlawful in that, is there?' The local flatfoots retired baffled, and the sign remained in place.

(*En passant*, Frank was anxious to trace the courses and the underground streams which ran through his land round the tower. Using a forked hazel twig, I found them, and I was right every time. Don't ask me why. Water divining shouldn't work – but it does.)

There are not many public observatories in Britain, but today many local astronomical societies have them. I have had countless visitors over the years, and I hope I have been able to help. For example, a long time ago a teenager tapped on my door and asked if he could see the telescopes. If I had said 'no', he might have left it at that. In fact he is now an extremely eminent professor of astronomy, and has left me far behind. He has contributed more than I ever could, so I am glad that I did not close the door and send him away.

20 The Weak Arm of the Law

I had my first (and only) brush with the Law a few years ago. I was driving down a country road, when a policeman shot out, stopped me and accused me of speeding. Technically I was; we were in a 30mph limit, and I was doing 35 to 40.

I have little sympathy for the careless driver, and none at all for the drunken driver. In fact, I would say that anyone who causes an accident after having had a pint too many should be banned for life. I have never driven after even one drink, because I know that I have slow physical reactions and it would be simply stupid to make them any slower. But on this occasion it was a clear, straight road, and I was little over the speed limit. The policeman was obviously out motorist-hunting, and he looked pretty gormless, so I decided to take a chance. 'May I see your warrant card, please?'

He wasn't carrying it; he was at my mercy – so he thought, at least. I gave him a good dressing down, told him that he was breaking the law, said I would put in a report to his Chief Constable (whose name I didn't give, because I didn't know it), and drove off. That was that. If he had been older and more experienced, he would have known that I was bluffing. (Incidentally, why do people talk about police *officers*? They do not hold Queen's Commission; they are not police officers – they are policemen.)

That episode started me thinking. I can remember the time

when we in Britain had the best train service, the best bus service, the best postal service, the best criminal justice system and, above all, the best police force in the world. Sadly, the situation today is not nearly so clear-cut.

When I first came to live in Selsey, in 1968, there was always a police presence even during the night; burglaries were rare, and vandalism almost non-existent. I used to work well into the early hours, and very often a policeman on patrol would tap on my study window and come in for coffee. The emphasis then shifted onto motorist-hunting, the night patrols ceased, and burglars flourished. It was said (whether correctly or not) that one senior police official had had a relative killed by a hit-and-run driver, and had become paranoid about motorists in general. In any case, the Selsey residents had finally had enough, and a petition with many hundreds of signatures was presented to Police HQ. It had absolutely no effect. I was told that police were actually available, the reason being that there were patrols in Chichester – and Selsey comes into that area. Yet there is an eight-mile road separating the two.

Bizarre cases have often come to light. One driver was prosecuted for taking a sip of water when stationary at a traffic light. Perhaps the best instance was that of a motorist who was taken to task for removing his hand from the driving wheel – but if that is a crime, how exactly do you change gear when driving a non-automatic car? The end result of all this is that the police are generally disliked, and this is unfair to those members of the force who have no choice but to obey orders from their superiors.

Mind you, the police are not helped by the attitude of some judges, who give the impression of being sympathetic to the attacker rather than the attacked. In my view at least there is an urgent need for a change in the law. Take the case of the Essex landowner whose property was invaded by 'travellers' who caused a great deal of damage. He unplugged the revellers' generator, and was busy arguing with them when the police arrived and arrested him. Even in AD 2002 I found this rather hard to believe, so I wrote to the Chief Constable of Essex asking him if the newspaper report could be correct. According to *The Times* of 4 January, a Police spokesman had said: 'We had just one police sergeant, 70 ravers and one unhappy landowner. The sergeant tried to calm the situation, but felt that it was getting out of hand. He felt violence would be caused and that public safety was threatened. In view of this he arrested the property owner.'

I was frankly surprised to receive a frank, honest reply from the Chief Constable, Mr. David Stevens, I am used to evasions, excuses and fudging. But Mr. Stevens wrote: 'Thank you for your letter of January 4, concerning the unlicensed public event which took place at Mountsales Farm, Dunmow, Essex, on the night of 26 December 2001. While I am unable to enter into discussions about the details of the case I can confirm that the facts as reported in *The Times* are broadly correct.'

Law and order?

To show that the country has gone raving mad, I quote the case of Mrs. Barber of Northampton, who is 93. She was burgled four times in 2002, so she arranged for barbed wire to be put around the top of her garden fence. She was officially ordered to remove it, in case it injured any burglar trying to break in. What can one say?

We also need to look at local magistrates, who are a very mixed bunch. Bossy women (invariably divorced) are common on the Bench, and have plenty of spare time, but it is true that anyone doing a proper, full-time job would not be able to give enough time. Therefore, your typical local magistrate must fall into one of five classes:

Too young to work. (I doubt whether this has ever happened).
Too old to work.
Too lazy to work.
Unemployable or
On the fiddle

It is hardly surprising, then, that the sentences passed are a little uneven. I take two cases reported in the papers in May 2001. One man was given three months' imprisonment for collecting used golf balls from lakes; another was given a conditional discharge after assaulting a milkman and putting him in hospital to be stitched up. It may or may not be significant that the first culprit was a white Englishman, while the second rejoiced in the name of Shahid Akram. Which brings me onto the two most pernicious Acts passed by Parliament in modern times: The Race Relations Act, and the Sex Discrimination Act.

Already I can hear the puerile squeals of 'Racist!' and 'Male chauvinist pig!' Well, if 'racist' means agreeing with Flanders and Swann that 'the English, the English, the English are best', then I am a racist. Unfortunately, the Politically Correct crackpots take it to mean something quite different, and this is where the trouble starts.

Initially, I doubt whether anyone can have been less 'colour conscious' than I was. I well remember walking down East Grinstead High Street when I was about eight, together with the local photographer and his son, Norman. A black man approached – a rare sight in those days – and Norman piped up: 'Look, Daddy, there goes a negative!' This was quite in order in 1931, but in 2003 the Politically Correct Brigade would have pounced upon Norman and charged him with 'racism'.

The Race Relations Act is the main cause of the trouble; it is driving a wedge between black and white, and it dominates the media, always with the same message: 'Black is Beautiful, White is Wicked.' Abolish the Act, stop regarding white Anglo-Saxons as second-class citizens, and the situation will solve itself. So far as immigration is concerned – the plain, unpalatable fact is that too great an influx is detrimental to the needs of our own people. This is not being 'racist', it is merely being honest.

Defending oneself has become illegal, and I once had an experience which is relevant. At around midnight I was collecting my car from the park outside the BBC when I saw two teenagers, both holding knives, bearing down on me with the clear intention of mugging me. Fortunately I was rather good at that sort of thing, so I laid them out. Call the police? No, no – they were black, and one had a broken arm. Once the police arrived I would have had to face the Race Relations Board, not to mention the Stephen Lawrence industry – so I drove off.

The Sex Discrimination Act? Well, again let us be honest. It works only one way, and often leads to third-rate women being promoted ahead of far abler men – which is bad for both sexes and moreover, bad for the country.

The trouble at the moment (2003) is that the Government has not the faintest idea what to do about all this, and there is to all intents and purposes no Opposition.

We must not give up. We may see the day when the police are rooted out of their panda cars, put back on the beat and told to concentrate upon crime rather than bullying motorists; when judges and magistrates are reminded that mugging an OAP is a more serious offence than selling bananas by the pound rather than the kilo; when immigration is restricted to people who, regardless of colour, are ready to adopt our ways and play a useful rôle in the society; when teachers go back to the old methods, and make sure that children leave school with the ability to read and write; when politicians tell the truth, at least occasionally, and when we break free from the dead hand of Europe and become once more a truly independent nation. Hitler and Mussolini could not beat us more than half a century ago; the only people who can beat us today are our own politicians.

21 Globe-trotting

Travel is great fun. I am no intrepid explorer who likes nothing better than to hack his way through the impenetrable jungle or perhaps sail a fragile craft up the reaches of the crocodile-infested Amazon, I am the first to admit that I like my creature comforts. Certainly I enjoy seeing the world, and by now I have been to most countries, mainly, though not exclusively, on astronomical errands. I have already said something about eclipse trips, and of course, I have been to most of the main observatories, but there have been occasions when I have travelled just for the sake of travelling.

I have missed out on India, and indeed the whole of the Indian sub-continent, because for some reason or other it has not appealed to me. Neither have I been to China, which I regret. I would like to see the Great Wall of China, around which a curious myth has grown up. It has been claimed that it is the only man-made object on the Earth which could be seen with the naked eye by an observer standing on the surface of the Moon. This is nonsense, the Wall can barely be made out from Earth orbit, and it is not continuous. Yet the story refuses to die.

One of the first forays after the war was to Iceland, which I liked then and still like now. Many people have wrong ideas about it, and believe it to be permanently dark and frozen. Actually it is not even in the Arctic Circle; the Circle just touches the island of

Grimsay, off the north coast, and there is one day in the year when from Grimsay the Sun does not rise (I gather that the islanders are rather proud about this). In high summer, of course, there is no proper darkness at all. The climate is chilly, but not intolerably so even in winter, when the skies are lit with superb displays of aurora borealis.

Reykjavík, the capital, is a cheerful place, and a lively one. I have always felt comfortable even though I cannot speak a word of Icelandic – which is an ancient language, not too unlike Anglo-Saxon, but with two extra letters, a crossed-out D (ð) and a p with a long tail (þ), both of which are pronounced *th.* Most people speak English, and all of them speak Danish, because Iceland came under the Danish crown before becoming totally independent in 1945. When I first went there I could just about stammer my way through in Danish, though I am ashamed to admit that I have forgotten it all now – I haven't tried to speak it for well over fifty years. The language has not changed much over the years, and I am told that the mediaeval sagas of Snorri Sturluson are written in what is virtually modern Icelandic.

Eating is great provided that you watch out for certain delicacies; the national dish (which tastes like poached walrus), rhubarb soup, very cold fried eggs straight out of the fridge, and ordinary potatoes with sugar on them. Of course raw fish is always on the menu; I happen to like it.

Iceland is purely volcanic, and there have been devastating eruptions in the past; much of the island is covered with lava, and trees are absent. In fact Iceland is infertile, and the economy depends on fish, which not so long ago led to a confrontation. Do you remember the Cod War?

Iceland decided to extend its territorial waters, and to ban foreign fishermen from coming too close in. This resulted in howls of rage from many countries, including Britain, and there were angry scenes. I am glad to say that the hostilities between Britain and Iceland resulted in no casualties whatsoever, and in the end it was Iceland which came out on top.

This was the only time when I was definitely on the side of our adversaries; the Icelanders depend solely upon fish, and I have every sympathy for them. I have one vivid memory of that episode. By sheer chance I was in Reykjavík at the time, and I was standing on the edge of the quay when an Icelandic naval officer came along; he saw the RAF tie that I was wearing (and which I still wear), and began chatting in the way that Servicemen do. I had dinner that night as a guest in the wardroom of the *Thor*, Iceland's leading combat vessel. Do you think that I ought to have been sent to the Tower?

One of Iceland's famous volcanoes is Hekla, which last erupted in 1947. I managed to scramble up its lower slopes, and it was something I would not have missed. With me was my second cousin, Brian Gulley (twenty years younger than me), but even Brian decided against trying to reach the summit; he would have managed it, but climbing is not my forte. We went to Thingvellir, seat of the oldest parliament in the world and one of the few which is still genuinely democratic; we sailed on Lake Mývátn, and discovered, when we were some way out, that our boat was leaking, and I did not like the idea of swimming in that icy water, though luckily we plugged the leak. We did bathe in one of the volcanic pools; it was like plunging into a hot bath. We sailed round the offshore island, right over the place where an active volcano, Surtsey, suddenly appeared a few years later; and we visited the falls of Gullfoss and Dettifoss, which are far more dramatic than Niagara.

One more memory of Iceland: I was with a leading geologist, Professor Sigurðar Thorarinsson, and from his study window we looked out over the lava. He had just come back from a series of conferences, and he had been to London, Paris, Athens, Rome and elsewhere. As we drank our coffee, he leaned back in his chair, and with a satisfied smile pointed out at the barren landscape. 'You know,' he said, 'it's so nice to get back to civilization!'

I knew just what he meant.

Norway is one of my favourite countries; it differs from Iceland inasmuch as a large part of it lies well within the Arctic Circle. Tromsø is a particularly attractive city, even though it does have a night lasting from late November through to late January. It is prosperous – indeed, all Norway is prosperous, and the economy is booming, because the Norwegians have had the sense to keep right out of Europe. You see magnificent displays of aurora from Tromsø, and I have taken parties there especially for photography; I recall one dear lady who wanted to image the aurora by using flash (how dim can one get?). The little fishing villages near North Cape are picturesque, but they must be bleak in the winter. Someone said that at such times there are only two things to do, and the other one is to play chess.

I was fascinated by the far North, it was only much later – in 1998 – that I was able to go to the far South. Believe me, it was worth waiting for. With me was Chris Doherty.

There were several reasons why I had always wanted to go to Antarctica. First, I like penguins, which are such delightfully improbable creatures. Secondly, not many people go there, at least so far, and there is no danger of overcrowding. And thirdly, there is something intriguing about a continent which is larger than Europe and yet has an indigenous population of 0. More or less out of the blue, I was invited to become a guest lecturer on a luxury cruiser, the *Marco Polo,* scheduled to make the great Antarctic crossing from Tierra del Fuego to New Zealand. My rôle was to give some talks about astronomy and to mingle generally, which sounded extremely pleasant, so I lost no time in accepting. I collected Chris and we set off.

We flew to Tierra del Fuego, at the southern tip of Chile, and then boarded the *Marco Polo* together with the rest of the party

– about 200 of us all told; a group larger than that would have been unwieldy. Naturally, we had kitted ourselves up with the right sort of gear. You need Antarctic clothing, because even if you go in February, which is the middle of southern summer, it becomes decidedly cold; you have to have several layers of clothing, plus waterproof boots if you want to go ashore (as we did). Not that there was any lack of comfort on the ship itself; far from it. The *Marco Polo* began life as a Russian icebreaker before being converted into a luxury cruise liner, so that she was ideally suited for polar journeys. Moreover, we had a Norwegian captain, and if you want safe, expert seamanship you can't beat a Norwegian.

Drake Sound, between Tierra de Fuego and the Antarctic Peninsula, can be rough, but on this occasion it was not and the sea remained calm. It was intriguing to note how the length of daylight increased; before long we were in the polar circle, under the midnight sun, and I had my first view of the Antarctic coast. There was something eerie about it, but it was the very reverse of unfriendly. There were huge icebergs, plus seals, whales and penguins, none of which seemed to be in the least alarmed.

My rôle was to give some talks about Antarctic astronomy; with permanent daylight we could not see the stars, but there was plenty to say. At present a major observatory is being built right at the South Pole, though of course we were limited to the coastal regions (travel inland is a very different matter, and strictly not for the tourist). As main speaker we had no less a person than Sir Edmund Hillary, who knows more about the polar regions than anybody else, and whose accounts of his past expeditions were riveting. Other lecturers dealt with the geology, the general terrain, the history of the whaling industry (now, thankfully, almost extinct) and the wildlife, which is unlike any other in the world. On board we had passengers of all ages and all kinds. One of them was Mrs. Finn ('Jackie') Rønne, who was the first woman to visit Antarctica, and after whom the Rønne Ice Shelf is named. Jackie once spent a whole winter ashore, and it must have been, literally, the experience of a lifetime. Her husband, sadly no longer with us, had been one of the great Antarctic explorers.

The weather remained good and the sea placid, and the cold was by no means forbidding provided that you took sensible precautions. Then came our first foray ashore. The procedure is to climb down the ship's ladder, jump into a circular rubber craft rather like an Eskimo kayak, and then be taken to the shore; the 'drivers' of these rubber craft are amazingly skilful. Once you are really close-in, you clamber out and wade ashore through icy, muddy water until you reach terra firma. Waterproof boots are essential; without them, the procedure would be most uncomfortable.

We duly landed on the Antarctic Peninsula, and I had my first close-range view of penguins – thousands of them. As I have said, they are improbable birds, and rather smaller than I had expected. They are not in the least nervous; after all, they have never been hunted, and are in no fear of human beings, so that they will waddle up and to all intents and purposes pose for you. We had been asked not to go too close to them, and everyone was careful to obey these sensible instructions. Once we even saw a king penguin, though in general the 'kings' do not frequent the coastal plains.

There is one problem which affects most people. A penguin rookery has a distinct effluvium – in other words, it pongs. I have absolutely no sense of smell, and so I was exempt, but others found it rather overpowering, particularly when wading through layers of penguin guano. The ground also tends to be slippery, and to land face-down on such a surface, as one luckless member of our party did, is emphatically not to be recommended.

We were free to walk around, and the more we saw the more fascinating it became; Chris took innumerable photographs, and

I even attempted some myself. On return to the ship we were helped on board by members of the crew; we took our boots off, and they were hosed down. They needed it.

On this voyage we were lucky. Going ashore depends upon what the weather is doing, and there are times when landing is impossible, but we were able to visit half a dozen sites, all of which were different. The main American base, McMurdo, was iced up, but we were able to go to the Italian base, and were made remarkably welcome by the scientists and crew members, all of whom went out of their way to make us feel 'at home'. Antarctica is a very valuable scientific site, and work goes on there all the time; even in the middle of the long, bitter night there are still scientists busy with their researches. I gather that the Italian base is a particular favourite for visiting scientists from other Antarctic stations, because of the renowned excellence of the Italian chef!

One amazing view was that of Mount Erebus, the world's southernmost active volcano. It was not erupting violently, as it sometimes does, but we could see steam issuing from the summit, and – unusually, so I was told – the whole volcano was cloud-free. Chris and other photographers made the most of our good luck. Incidentally, electronic devices were liable to freeze up in the Antarctic, and we had prudently equipped ourselves with purely 'mechanical' cameras, though in the event the temperature was not so low as it generally is at that time of the year.

We went to Shackleton's base, and then to Scott's hut, where Captain Scott and his companions spent their final winter before their fateful journey to the South Pole. There was something uncannily poignant about the place, because it looks so like an ordinary, undistinguished hut – until you go inside. It was tragic that the explorers were so close to safety, and yet could not make it – and here we were, less than a hundred years later, travelling in comfort and safety. Also, it does seem ironical that although everyone remembers Scott, far fewer people remember Roald Amundsen, who got everything right and was the first man to reach the Pole itself. At least, during our journey in the *Marco Polo,* we sailed right across the Amundsen Sea.

We worked our way along the coast, going ashore wherever we could, there was always something new to see – and photograph. Even the penguins were different from one site to another. We saw whales aplenty, and one seal almost waddled up to us, as though deliberately posing and saying 'Here I am! Photograph me!' Then, too, there were the glorious colours associated with the midnight sun; I had seen the same sort of thing from the Arctic, but nothing can equal the beauty and the grandeur of the far south.

Finally, it was time to go. As we stood on the deck of the *Marco Polo* and had our last views of the great white land fading into the distance, we somehow had the feeling that we had been visiting an alien planet. 'There's nothing like Antarctica,' said Chris, and I agreed; once you have been there you have a more complete idea of the world. I am sad that I now know that I will never to be able to go back.

There was one final flurry of excitement. On the way to New Zealand the weather, which had been so kind, changed abruptly, and in the Tasman Sea, we ran into a storm so violent that even the robust *Marco Polo* was thrown around. One freak wave went so far as to damage a lifeboat. At the time I was playing bridge in the main lounge, partnering Jackie Rønne, and the jolt caused by the wave was accompanied by a noise like a thunderclap. I will always maintain that this is what made us call a totally unbiddable Six Clubs – and go four down.

We came home via New Zealand, which is another of my favourite places; life there is less hectic than elsewhere. I remember being caught in a traffic jam in Auckland which held me up for at least three minutes, but that was exceptional. At one time or another I have seen a good deal of both islands, and each has its own special attractions, from the beauty of the Southern Alps to the vulcanism of Rotorua. White Mountain is New Zealand's most famous volcano, and it is mildly active all the time.

On one visit, some years ago, a friend flew me close to the summit in his light aircraft, and we were well and truly thrown around, so that my photographs did not come out at all satisfactorily.

There are two major differences between New Zealand and Australia. First, Australia is plagued with flies; they are everywhere, and you simply cannot get away from them. In New Zealand there is almost a complete lack of them. I asked why, and was given a straight answer: dung-beetles. These beetles eat the dung, and prevent the flies from spreading, whereas Australia simply does not have enough dung-beetles to do the job. I have wondered why the Australians don't import dung-beetles by the cartload, but no doubt there is a good reason.

The second difference is that New Zealand has no 'nasties' lying in wait, while in Australia there are all manner of dangerous creatures, notably almost every variety of snake and some most unpleasant funnel-web spiders. This is equally true of the sea. Bathing in New Zealand is safe (apart from the very occasional wandering shark), but this is not true of Australia. I was once on the beach at Darwin, looking out over a calm, beautiful sea, positively inviting me to go in and have a swim. I didn't, and this is just as well, because I might not have come out alive. The water contains large numbers of what are called sea wasps, but are in fact jellyfish; tread on one of these, and you will be dead in a matter of minutes. Off New Zealand, the sea wasps are not there.

I have been to Australia often enough, and I have enjoyed it. To me, one very fascinating place is Wolf Creek Crater, near the boundary of Northern Territory. This is an impact crater, caused by a meteorite which landed there thousands of years ago. It is well-formed, and climbing up the wall was easy even for me, but the best way to see it is to fly over it, which is what I did.

Impact structures are commoner than used to be thought, but many of them have been so eroded that they are difficult to recognise. Wolf Creek has escaped this fate, and so has the most celebrated of all impact structures, Meteor Crater in Arizona, once described by the eminent Swedish scientist Svante Arrhenius as 'the most interesting place on Earth'.

I had my first view of it in 1954, when I had been to the Lowell Observatory on one of my Moon-mapping trips. I was flying from Place A to Place B on one of the small airlines which operated in those days, but which have long since been swallowed up. I was the only passenger – the plane was a six-seater – and seeing my RAF tie the pilot invited me to join him, so I settled comfortably into the co-pilot's seat. It occurred to me that if we made a slight detour we would go right over Meteor Crater, and the pilot was quite happy to do so. We were lucky, because it was a fine day and the air was calm; we made a couple of passes, and I took photographs. Then the pilot had an idea. 'I've never been over this thing, or at least I haven't taken any notice of it. Will you fly the kite while I get my camera out?'

This was no problem; I was used to flying small, light aircraft, and I felt quite at ease. We circled the crater while the pilot took his photographs, and then flew back. Of course there were two sets of controls, so that if I had shown signs of insanity the pilot could have taken over – but imagine what the owners of a modern airline would say if a passenger was allowed even to touch the controls? I last flew solo in 1945, and I am, I fear, a little rusty.

You can reach Meteor Crater easily; it lies not far from the town of Winslow, and all you have to do is to drive along Highway 99 until turning off on to the crater road. You may not recognize the crater until you are almost there, because although it is nearly 600 feet deep there is not much in the nature of a wall rising above the desert. Like all impact craters, on the Earth and also on the Moon, it is a sunken formation.

The name is wrong. It really should be Metor*ite* Crater, because it was produced not by a meteor of the shooting-star variety, but by a large mass of iron-rich material which swung around the Solar System for thousands of millions of years before colliding with Planet Earth, about 50,000 years ago. There was nobody around to watch it fall. When *Homo sapiens* came to Arizona, Meteor Crater was already ancient, which is why the various legends about it must be discounted.

The crater is so well defined because the land is exceptionally well suited to preserving a huge scare; there is not much in the way of erosion, though no doubt the crater has been filled to some extent by windblown material, and is less deep than it was originally. Moreover, human activity has been at a low level. There has been only one attempt at serious interference with the crater, and this piece of scientific vandalism was halted before much damage had been done; Meteor Crater had the last word.

Because the crater is so isolated, it was not known to white men until less than a century and a half ago. Even when it was first reported, in 1871, nobody took much notice of it until the arrival of an American geologist, G. K. Gilbert, who was interested in it, surveyed it, wrote an account of it – and came to a completely wrong conclusion; he believed that the crater was 'a steam explosion of volcanic origin'.

Next came Daniel Moreau Barringer, who plays a dual rôle of hero villain. He looked at the crater, and as a mining engineer he was interested, though his reasons were very different from Gilbert's. He seems to have realized quite quickly that the crater had been gouged out by a meteorite, and he knew that meteorites can be valuable commercially; some are stones, but others are made up of excellent-quality iron, and there are appreciable amounts of more exotic elements such as iridium. So in his hero's rôle, Barringer bought the site and preserved it for posterity. In his villain's rôle, he began to work out how he could fish the meteorite up. Clearly it would be too large to be brought to the surface intact, but if it could be reached it could be chipped away.

Barringer brought down mining gear, and started to look. He had no luck, and in fact he made a major mistake; he believed that the meteorite must be buried below the centre of the crater. This seems axiomatic, but impact craters are less straightforward than might be thought, because a violent impactor will produce a circular crater even if it comes down at an angle. When the missile hits, its kinetic energy is converted into heat, and acts as a very powerful explosive; as soon as the meteorite has buried itself, the explosion blasts out a crater which is basically circular. (All the lunar craters are like this, though many of them have been badly distorted by later impacts.)

Barringer also found that the floor material was unexpectedly hard; there is sandstone down to an appreciable depth, and then a layer of which is called kaibab dolomite, which is really tough. Then he switched his attention to other areas, with no better result. Finally, at a depth of less than 2,000 feet, the drill jammed. Nothing more could be done, and Barringer gave up. A few later sporadic efforts were just as fruitless, and all that is left today is a remnant of the old mining gear, which is frankly an eyesore.

Later still, it was found that the silica round the crater rim is of very high quality, and again the ghouls of Big Business showed interest. There was a considerable amount of activity, and the silica miners became an unmitigated nuisance, but this time officialdom stepped in, for once on the right side. All work on excavating or collecting was stopped permanently, and in 1967 the crater was declared a National Natural Landmark. A Board of Administration was set up, and a museum was established on the crater wall, just below the rim: It contains specimens of meteorites found in the area, as well as detailed explanations of how the crater was formed. We are used to regarding Government officials as pests, but in this case they acted very properly. Few people will object to paying a modest entrance fee.

On my first 'ground' trip, in 1955, you could walk down the 'Trail' to the crater floor, and I did so; it took some time to go down and longer to come back, under a broiling Arizonan sun. Today the Trail has been closed, because it is unquestionably dangerous, and apparently there was a bad accident some years ago. Not that this matters to the average visitor; the best vantage point is the crest of the wall, and it does not take too long to walk all the way round, because the crater is less than a mile in diameter.

I last went there to present a *Sky at Night* programme. We had helicopters to take our television equipment down to the floor. For one shot Pieter Morpugo wanted to film the scene in the late evening, so I was flown down and dumped. Darkness was falling, and to be totally alone there, cut off from the rest of the world, was quite an experience; I have never felt so isolated. To the best of my knowledge, nobody has ever taken up residence to the crater. It would lead to a decidedly limited outlook.

During that trip we spent some time in or near the crater, and I remember a remark made to me by Jim Greenacre, one of the Lowell astronomers, while we were walking around in the undergrowth near the edge of the rim. 'Haven't seen one rattlesnake today,' he said cheerfully. I wasn't sorry; you can put me down now as a coward.

I have seen a good deal of North America, and there is endless variety. One State I like very much is Alaska. I went there with a party of astronomical enthusiasts with the aim of photographing aurorae, which are well seen from there even though I personally think that North Norway is better. We flew to the capital, Anchorage, and then went on to Fairbanks, which is pleasant though bitterly cold. On the train journey back to Anchorage when a herd of moose showed up. The train slowed down, and the conductor came into the carriage. 'Any of you want to take photos?'

We did – and the train stopped, giving us ample time to collect our cameras, disembark and photograph the beasts which were obliging enough to pose for us. It was a nice gesture on the part of the train driver and conductor, and indeed we found that the Alaskans are among the friendliest people in the United States.

South America is as different as it could be. You have to be on the alert all the time, as I found out in Argentina during the IAU General Assembly; the only stable country seems to be Chile. Whatever one thinks about General Pinochet, there is no doubt that he restored law and order, and left Chile in good shape. I hold no brief for his methods, but it is worth recalling that he was friendly to us during the Falklands war; we repaid him by kidnapping him when he came to England for medical help, and held him for months. Yet we put out the red carpet for Robert Mugabe of Zimbabwe, who must be one of the most unpleasant people on the face of the earth...

I went on to Colombia, my mission being to act as an IAU representative at the inauguration of the Colombian Astronomical Society. The capital, Bogotá, masquerades as an open sewer; the Oberservatory is in Government grounds, and is strictly quarantined when a political coup is going on, though during my stay there was no unrest (in fact, I understand that there were only two revolutions in the entire year). The next stop was Medellín, Colombia's second city, and recognized as the drug capital of the world. Landing at the airport is an experience, because Medellín is mountain-ringed; the plane seems to aim for the centre of the peak and then bounce up over the crest on to the airstrip. I was none too sure about Colombian pilots, though I have to admit that they appeared to be quite competent.

Outwardly Medellín looks normal enough, but I was conscious of a curious atmosphere even though I cannot speak Spanish. Everyone was friendly, and I hope that I carried out my duties in the proper manner, but I would not be happy about staying there for very long. I am told that the smell of drugs is unmistakable, though of course it was lost on me.

My other interesting spell in South America was in 1986, when I went to the launch of the Giotto spacecraft to Halley's Comet. Our television programme was not a success, as I said earlier, and although I was at the launch – at the request of British Aerospace, who wanted me to do some commentaries – I was unable to come on to the BBC 'live', the cost of sending a television unit was said to be too high (!). I did, however, cover the launch for television in several European countries – I am not now sure which.

Remember, Giotto was a 'one-off'. It had to work, faultlessly, first time, otherwise there would be a delay until the comet came back in 2061. The fact that the Americans were not involved meant that the launcher had to be a European rocket, and the only one available was a French Ariane. It was true that the Arianes had had their problems, but there was no alternative.

The launching site was at Kourou, in French Guiana. I flew to Paris and then went on to a Concorde – my first flight in this remarkable aircraft. I am not going to say that it is roomy, or more comfortable than an ordinary aircraft, but it is certainly fast, and the take-off gives the impression of being almost vertical. The Transatlantic flight gives you just about enough time to down a couple of gin-and-tonics.

We had a couple of days in Kourou before the launch, and were able to look around. I was able to visit Devil's Island, which has the reputation of being just about the nastiest penal settlement of modern times unless you count the German concentration camps. With a party of astronomers I went over by boat; and halfway there we were surprised to see a man apparently surfing. Asked if he needed any assistance, he waved airily and told us that he was quite all right. I assume that he knew about sharks.

Devil's Island is calm and peaceful now, but the old buildings are still standing, and there is an undeniable atmosphere of menace which I found disturbing. Some of the ex-convicts were still around; one of them kept a café in Kourou, and we had lunch there before the Giotto launch.

The launch itself went off without a hitch. It was a relief to see Giotto safely on its way, I wondered what it would find when it met the comet.

Other places? Well, I have been to Borneo (no cannibals) and to Vietnam (which I liked), each time during an eclipse trip with Chris. I went to Malaysia, because I was asked to give an opinion as to a site for a proposed Observatory in Kuala Lumpur, I am glad to say that my recommendation was followed. While in K. L. I decided to buy a new sports jacket; these can be made by the local tailors at a moment's notice, and they always fit. I chose what I thought was an Oxford blue, and emerged into the daylight – the shop itself was extremely dark. When I collected the coat next day, it turned out to be the wildest electric blue you have ever seen. I keep it for parties!

Finally, in this rambling travelogue: Egypt. I had always wanted to see the Pyramids, and eventually I did; they are impressive, even though they do lie in the garden of a hotel. We went to the Valley of the Kings, which did fit with my ideas about Tutankhamen and Ramses, and at one point Chris and I were standing in the Valley of the Queens – the precise point where, twenty-four hours later, a gunman opened fire on bystanders and killed several. We were lucky.

I remember Luxor mainly for its traffic jams, and Cairo for its noise. Most of our party had problems after eating hotel food, but we escaped, because we kept entirely to oranges, bananas, bottled water and neat whisky. All in all Egypt is a fascinating place, even if it is rather a pity about the Egyptians.

If I had the chance I would go to the South Seas, Tahiti in particular, and I would like to look at Easter Island, with its mysterious statues. Alas, my travelling days are over, but I have the satisfaction of knowing that I have been able to see more of the world than most people have had the opportunity to do. I certainly can't grumble. Yet no matter how far away I have been, I'm always glad to drive back along that road from Chichester, with its fifty-three sharp corners, knowing that Selsey lies ahead.

22 The Tale of Mr. Twitmarsh

In recent years I have become embroiled with two sets of curious people: Twitmarches, and the Thought Police.

Have you ever heard of a Twitmarsh? Possibly not, because it is a name that I invented myself. It honours Mr. K. Whitmarsh, of the Southern Gas Board, who first came into my life some years ago, when I had lived in Selsey long enough to have become part of the local scenery. I was firmly rooted in Farthings, my old, thatched house, which is my home and which I would never leave.

One glorious spring morning Mr. Whitmarsh sent me a Final Demand. If I wanted to escape the full rigour of the Law, I must pay £10 for repairs to the central heating system in Farthings. As this is powered entirely by oil, the situation was fraught with interest at once. I could of course have told Mr. Whitmarsh to go fly a kite, but I rather wanted to see what would happen, so I sent a cheque for £10 (which actually couldn't be cashed, because I had dated it wrongly) together with a mild inquiry. In successive days I then had:

A letter from Mr. Whitmarsh, saying that I probably didn't owe the £10.

A second letter from Mr. Whitmarsh, saying that I certainly didn't owe £10.

A refund of my £10.

A second refund of my £10.

A second Final Demand.

I checked up, and found that other people had heard from Mr. Whitmarsh in the same vein. I christened him Twitmarsh, which seemed appropriate. And I did some thinking. What if this Final Demand had landed on the doorstep of a nervous pensioner? And in any case, why should the great British public be harassed by Twitmarshes who are safe in their Civil Service funk-holes, have immunity from dismissal, are sure of nice, comfortable pensions and can hide behind a cloak of anonymity? Time to go on the offensive. So I decided to have some fun with bureaucrats in general, and created the author R. T. Fishall (as a pseudonym; I considered R. Hugh Hall-Wright, but it seemed a little too obvious). The first Fishall book – *Bureaucrats: How to Annoy Them* – went well, and as I write these words both it and its successor, *The Twitmarsh File*, are being reprinted.

Meantime, I had had another Final Demand, this time from the North Thames Gas Company, calling in £865.47 for installation of gas central heating in a house in Lowndes, Street, S.W.1. This one was signed 'W. Bonney' – not 'pp W. Bonney', please note. I found that the signature had actually been written by a Miss Whitty, of the same department, so that it was technically a forgery. There really was a Moore who owed a bill, and Miss Whitty had simply sent the demand to the first Moore she found in *Who's Who* … Me.

After some difficulty I got through on the telephone to Mr. Gadd, the Chairman of North Thames Gas. The conversation went like this:

Me. Well, what is the explanation?

Mr. Gadd. It was a mistake in our computer department.

Me. Oh no, it wasn't. It was a slip-up by your Miss Whitty. Didn't you know?

Mr. Gadd. Gulp.

I then established that the real owner of the Lowndes Street house had departed for Spain, and whether Mr. Gadd and Miss Whitty ever caught up with him I do not know. At least he had launched my anti-Twitmarsh campaign, and I will give a few examples of my methods:

Fundamental Laws

When writing to a Twitmarsh, never be explicit. Make your letter long, verbose and only semi-legible.

Never give a correct reference. For example, if you have a letter from a Twitmarsh with reference EH/4/PNG/H8, head your letter back WS/3/JGH/H9. Also, request a reply to a letter that you haven't actually written, so that the Twitmarsh will spend time searching for a letter which has never existed.

In paying a bill sent by a Twitmarsh, make the amount slightly wrong – by a few pence, or up to a pound. If you have sent slightly too much, then ask for a refund. Note too that a well-placed staple in the middle of a cheque, attaching it to the accompanying letter, has been known to jam a computer completely. Crumpling a letter or cheque and then smoothing it out can have a similar effect.

If the Twitmarsh has not sent you a stamped addressed envelope, reply without adding a stamp – write OFFICIAL where the stamp ought to go. (*En Passant.* I am reminded of the man who persisted in putting the stamp in the exact centre of the envelope. Finally the Twitmarsh became exasperated, and wrote to ask the writer to be more conventional in future. In return he received a letter with the stamp central, as before. Inside was the message:

Hey diddle, diddle,
The stamp's in the middle.

I think he then admitted defeat).

Of course, these procedures should be applied only for Twitmarshes. I have not the slightest sympathy for anyone who owes a bill to, say, the local grocer and does not pay it promptly and correctly. This is basic honesty, and anyone who disagrees is not on my wavelength at all.

Twitmarshes also have their own jargon. For example:

Official Letter	*Translation*
Your letter has been carefully considered, and its contents noted.	I haven't looked at it.
A full survey of the problem has been put in hand.	Nothing will be done.
I assure you that action in this instance will be taken as soon as possible.	Nothing will be done.
I fully appreciate the problem.	I couldn't care less.
Your complaint is being fully investigated.	Your letter has been put into the wastepaper basket.
I do not really feel that any useful purpose would be served by pursuing this matter further.	Get stuffed.
I assure you of our attention and consideration at all times.	Now I'll pull the other one.
The problem to which you refer is being investigated, but you will appreciate that our department is extremely busy.	I'm playing golf at half-past three.

With care, Twitmarshes can be curbed, particularly since in general they lack a sense of humour, making them vulnerable. (I may add that I have met only three people who were utterly devoid of the faintest spark of humour; my paternal grandmother, a former President of East Germany (whom I regarded as the archetypal Sauer Kraut), and General de Gaulle. Of these, my grandmother had every excuse, because she was French-Swiss). Also humourless are the Thought Police, who really do have a malign influence. This brings me on to the curse of Political Correctness.

We are used to the term now, and it is ever-present in the Press, on radio and on television, but this was not always so. Originally it was a joke, and was confined to folk such as ardent feminists, but then the Thought Police moved in, and by now they have become a real threat. Ironically, one of the first targets was W. S. Gilbert. In the Mikado's opening song, the phrase 'blacked like a nigger' has been sanitised to 'painted with vigour'. When I first heard this I laughed, and I am sure Gilbert would have laughed too, but it was an ominous foretaste of things to come, and by now the situation has become ludicrous.

'Women's lib' looms large on the horizon. There is outrage about the use of the term *mankind* for the human species; the crackpots insist upon *humankind*, and in so doing they show their lack of education, because the official term for humanity is *Homo sapiens*. Many organizations now insist that there must be a definite percentage of women in senior positions – which is tantamount to saying that women must have special treatment because so few of them are capable of being promoted on their own merits. (Come to that, why not have special treatment also for left-handers, or people with red hair?) Bear in mind that no groups of people object to being regarded as inferior unless they really *are* inferior.

Sex? Well, anyone who now dares to suggest that homosexuality is undesirable is regarded as Politically Incorrect, regardless of the fact that homosexuals are mainly responsible for the spread of Aids (the Garden of Eden was the home of Adam and Eve, not Adam and Steve). While on the subject of religion, did you know that there have been serious attempts to do away with the terms BC and AD, because they may offend people of other religions such as Mohammedanism, Hindu, Urdu, or Voodoo? At a

Christmas nativity play put on at St. John's College, Oxford, the Three Kings of Orient became Three Queens.

The Bible itself has not escaped censure. In an official Church of England publication issued in April 2002 (!) at the diocese of Ripon and Leeds, all references to God as being masculine have been altered – presumably the official Church idea of God is that she's black. The Oxford University Press joined in the fun by producing a new version of the Psalms and the New Testament in which every reference to 'the right hand of God' was deleted because it might upset left-handers, while the Lord's Prayer began 'Our Father-and-Mother which art in Heaven…' When I first heard about this I naturally suspected a hoax, and wrote to ask. Miss Caroline Pailing, of OUP, wrote back stressing that it was not a hoax, and had indeed 'been welcomed by many Christian groups'. At that point I admitted defeat. And in 1998 the Birmingham City Council renamed its programme of Christmas festivities 'Winterval', in deference to non-Christians, while Christmas lights became 'festive lights'.

Another new psalm book, officially sanctioned by the Church of England, contains some sections which are open to misinterpretation. Thus King David's Psalm 6 becomes 'Every night I drench my bedding and flood my bedding'; while Psalm 107 refers to 'those who go down to the sea in ships and do their business in great waters'. No comment!

The Thought Police have invaded the whole of society by now. You must not call anyone 'a good egg', because eggs are associated with egg-and-spoon races, and 'spoon' rhymes with 'coon'. The word 'British' is now frowned upon by the BBC, because it may offend the Scots, the Welsh and others. The last night of the Proms in 2002 lacked the traditional Rule Britannia, which was axed as being militaristic, and in any case patriotism is not acceptable in a multiracial society. In Leicester, a Mrs. Nancy Bennett had her display of china pigs seized by the police, because it was 'offensive' to the Muslim neighbours. Stockport College, in

Greater Manchester, bans the phrase 'slaving over a hot stove?' because it minimises the horror of the slave trade, and also 'a black look', while a Chairman must become a Chairperson ... One could go on and on. Astronomically, I wonder when the crackpots will start worrying about white dwarfs, red giants and, above all, black holes?

The time has come to take stock and realize what these people are trying to do. The PC campaign has gone beyond all reason; it is quite definitely a deliberate attempt to undermine everything that we have fought for, and defended, for the past thousand years. Unless we stand up to be counted, the Thought Police will win where Napoleon, the Kaiser, Adolf Hitler and others have so signally failed.

23 The Power of the Press

Is there anything in your life which returns to haunt you even after many years? I can recall one instance: a snatch of conversation overhead in 1941, and which has baffled me to this day.

Two of us were in an open car on a hot summer day, driving down Oxford Street (you could do that in 1941). My companion, one of the closest friends I ever had or could ever have, was a senior RAF officer, and it was he who was driving; we were on our way to an official meeting. Down the road, on the far side, came another open car, chauffeur-driven – unusual at that time – and carrying two very aristocratic-looking ladies. As we passed, a voice floated over to us in a cut-glass accent. 'So, my dear, all I could do was to use the baby as a hammer.'

There was a stunned pause. 'Tommy,' I said, 'I've got to know.'

'Kid, so have I. Hold on.'

We swung round – too late; we had lost them, and we never saw them again; they must have turned on to a side-road. What on earth could it mean? That remark comes back to me even now, after more than sixty years. A real baby? Some sort of mechanical device? I will never know. If you have the sense of humour of an average ten-year-old – and to this I plead guilty – it is the sort of thing that can't be forgotten.

(Around the same time I made a real 'put-down' remark. I was not in uniform, and I went into a bar and ordered a whisky. The conversation went like this:

Barman: You, lad? You're not old enough to drink. Mummy wouldn't like it.

Me: (Haughtily producing my identity card). I happen to be an officer in His Majesty's forces.

Barman: (abashed). Sorry, sir. Large or Small?

Actually, we were both right. I *was* an officer in HM Forces. And I *wasn't* twenty-one).

Later on I began making a hobby out of collecting misprints and oddities, and by now have books of them, mainly from the daily Press. Basically, I have three sections – Headlines, Funny Peculiar and Funny Ha-ha – and it may amuse you to peruse a modest selection. Headlines first, and my favourite, from the *Sun* newspaper, is I fear a little unkind; the date was November 1995, a short-sighted man went into the bathroom to apply some haemorrhoid cream, but picked up the super-glue tube by mistake. The *Sun* summed up the situation neatly: OUR JOHN HAS GONE POTTY AND GUMMED UP HIS BOTTY. Luckily I gather that there was no lasting damage!

Here are some more:

ITALY'S FROG PUTS PRINCESS ON HOLD (*The Sunday Times*). Disappointingly this turned out to refer to an argument between an Italian diplomat and his wife.

PROFANE PARROT UPSTAGES ADMIRAL. (*The Times*, 22 June 2001). Sammy, the grey parrot mascot of HMS *Lancaster*, has a rich vocabulary. He went 'in the air' during a speech by the commanding officer, Sir Alan West, using terms such as 'S**t!' 'C**p!' and 'P***s!' He was then temporarily banished to his own quarters.

TUSSAUDS WAXWORKS URGED TO WALK OUT. (*The Times* again, May 2002). I would have loved to have seen this, particularly as I am standing there at the door leading to the London Planetarium. I have to admit that the model is so good that people can't tell which is which.

BUM GOES POP (*Buenos Aires Herald*, May 1994). This refers to a dispute between the Argentine Government and the country's main steel union, but I feel that it might have been more delicately put).

I will merely give you some of the others, and let you draw your own conclusions:

BAN ON SERVING ALCOHOL TO CUSTOMERS IN BOTTLES. (*The Scotsman*, October 1994).

RABBIT RUNS OFF WITH £25,000. (*The Times*, 27 June 2000).

ARSONIST GETS TIME TO FINISH MACBETH. (*Daily Mirror*, April 2002).

SECOND COMING TO BE BROADCAST ON THE INTERNET. (*Sunday Telegraph*, 11 December 2001).

MARCHWOOD TOILET BLOCKED BY PLANNERS. (*Lymington Times*, 1998).

MYSTERY OF THE DONKEY THAT BRAYS BACKWARDS (it says haw-hee). (*Daily Telegraph*, 15 February 2000).
and finally:

CHEFS MAKE AN ASS FIT FOR A PRESIDENT. (*The Times*, 6 August 2001). Please don't ask me what this is about; I have mislaid the rest of the cutting!

Turn now to some examples of Funny Peculiar:

Mr. Neil Symmons, of Stockeinteignhead in Devon, is an expert bird-watcher, and he made a habit of going outdoors at night and hooting like an owl, hoping that an owl would hear him and reply. After a while he received one, and after that there was a nightly tryst; it even seemed possible that some sort of real communication could be established. However, his neighbour, Mr. Fred Carnes, had had exactly the same idea. It was only when the two wives were chatting, months later, that Mr. Symmons and Mr. Carnes realized that they had been hooting at each other. What the owls thought about it is not on record.

In 1998 the first woman to captain a Royal Navy ship took exactly thirteen days to steer her boat into another boat. I am glad to say that there was no serious damage.

I come next to an event on 19 January 2000, widely reported in the Press. Police authorities had trained Clyde, a pointer dog, to sniff out hidden drugs. At one inspection of suspects he concentrated upon an Asian gentleman, who volubly protested his innocence; an official investigation was put in hand, and Clyde was condemned as a racist – hence a headline in the *Sun*: 'Dog on Racist Charge.' I feel that this was grossly unfair, but I doubt if Clyde was at all worried.

In Chichester, in 1997 the main cesspool firm changed hands. The buyer was a Mr. Pee.

The island of Little Cayman, in the Caribbean, is apparently a beautiful place (I have never been there). It has a traffic accident rate of 100 per cent. There are only two cars on the island, and one day in 1995 they met head-on.

There are some instances when a major police investigation becomes essential. This happened in 2000 in the Belfast headquarters of the Royal Ulster Constabulary. The biscuit barrel in the officers' mess was being consistently raided, and it became horribly clear that a dastardly Kit Kat thief was at large. Cameras were set up, and everyone coming in or out was recorded on film at the express wish of the officer in charge, Chief Superintendent Adrian Ringland. After a week the culprit was caught – a senior officer on the station, who had been shamelessly helping himself to Kit Kats without putting any money in the till. Of course, he had to be posted to another station, but Mr. Ringland was noticeably reticent about giving interviews about it all.

In 2000 the Kingdom of Swaziland, in Southern Africa, was thrown into turmoil when the Speaker of the Swazi Parliament was dismissed for stealing a piece of cow dung from the King's cattle enclosure.

A Mr. Brian Cheeseman threatened legal action after being ejected from Lord's Cricket Ground during a Test Match between England and Australia. At the time he was dressed as a carrot.

In November 2000 firefighters were called in at Chessington in Surrey, because an animal was apparently stuck at the top of a tall tree. Nobly they brought their ladders, erected them, climbed up, and rescued – a squirrel.

You're not going to believe this one: Lambeth Council wrote to Mr. W. Reynolds, saying 'Your council tax benefit has been stopped from 17 April 2000 because there has been a change in your circumstances. The change is because you are dead. If you think you may still qualify you must contact this office immediately to obtain a new application form.' Mr. Reynolds had indeed departed this earthly plane, but I suppose he could have engaged the services of a powerful medium.

A butcher's shop in Marlborough put up a helpful sign, 'Fresh Lamb. Mint Condition'. I don't know whether it is still there. One sign which has vanished used to be prominent at Surbiton in Surrey, at the entrance to a footpath: 'To Cemetery and Crematorium. Pedestrians Only.'

I loved a notice in the Torossy Castle Café, in the Isle of Mull: 'Will ladies please rinse out teapots and then stand upside-down in the sink. On no account must hot bottoms be placed on the worktops.'

The Watton-in Stome Women's Institute advertised a talk, 'Three Pairs of Knickers'. The speaker was a Miss I. Leek. And the *Hounslow Observer* carried an interesting veterinary advertisement: 'Low cost neutering service for owners whose pets are on low incomes.' I think I know what it meant...

How's this for a piece of humbug? Peterborough Cathedral has applied for a grant of £1.5m from the National Lottery, even though the Church of England bishops oppose the Lottery and even denounce it as immoral. The Churchdown Parish Magazine of April 1995 carried an important announcement: 'Will the congregation please note that the bowl at the back of the Church, labelled "For the Sick", is for monetary donations only.'

A shoplifter stole a lobster and hid it in his pants. Alas, the lobster became cross, and used its pincers in a way which has probably stopped him from becoming a father. And according to a report in the *Huddersfield District Chronicle*, Kirkbarton parish councillor Gerard Moore has taken over 'the fight to stop the abattoir from giving off smells from former parish councillor Rowena Ogle.'

In Pen-y-fal in Mid Glamorgan, the Council has taken a firm line with three pet ducks, Snowy, Mary and Quackers. They are officially banned from using the village pond, because the pond is reserved strictly for wildlife. Well – at least somebody must be quackers...

Education also comes under scrutiny. The headmistress of a school in London did not allow the children to see the ballet *Romeo and Juliet*, because this is 'a blatantly heterosexual love story'. As I doubt whether anyone can beat that, let me turn to misprints, some of which are hilarious. The *Lincolnshire Echo* recently featured 'a cuddly toy, a large guerrilla' while according to the *Belfast Telegraph* 'the rain came down in steroids'.

One could go on and on, but I am sure you will get the general idea, so I will merely cite what has sometimes been referred to as 'the misprint to end all misprints'. Long ago a new bridge over the Thames was opened by King George V and Queen Mary. *The Times* report should have read 'After the ceremony, Their Majesties *passed* over the bridge.'

To conclude – another favourite of mine. During a rugger match in Cola, in Western Samoa, a member of the Great Britain team was sent off the field. He was fined one pig.

24 Apollo – and After

Do you remember where you were on the evening of 20 July, 1969? I know exactly where I was: in Studio 7, BBC Television Centre, carrying out a commentary as Neil Armstrong and Buzz Aldrin became the first men to set foot on the surface of the Moon.

It was an epic moment – surely one of the greatest in human history. Now, over thirty years later, I am staggered to realize that there are some people who believe that the lunar landings never happened, and that the whole Apollo programme was faked by NASA. The idea came from a science fiction film, *Capricorn One*, with which NASA was sporting enough to collaborate, but nobody in their senses expected this sort of thing to be taken seriously. It is about as logical as the older idea that the Moon is made of green cheese, and how anyone can swallow it (the idea, not the cheese) passeth all comprehension. (However, I remember that during one of the Apollo broadcasts I solemnly rode the Emmet lunar bicycle; which really was equipped with a cheese comparator and a meteorite umbrella. One perfectly sincere viewer wrote to me and asked if that cycle had really been to the Moon!)

Throughout the Apollo period I was wearing two hats. One was my NASA hat, because I had become one of the accepted Moon-mappers, even if a very minor one. Secondly, I was one of the two main BBC commentators, so that all through the Apollo missions I was more or less a permanent resident in Television Centre. It was a never-to-be-forgotten experience, and we were all kept busy. We had many late-night programmes; I recall that every bar and restaurant in the building closed down at breakfast-time!

The original BBC plan had been to use James Burke as the main commentator, with me as the astronomer. In the event it didn't work out quite like that, and we shared the commentary equally, though our styles were rather different; James talked more than I did, because when action was taking place I preferred to say as little as possible. This is not a criticism of James; he was an expert broadcaster, and it all went very smoothly. In fact James was not trained as a scientist – his main degree was in Italian – but he was amazingly versatile, and when it came to describing rocket 'hardware' he was much better than I was. I am always sorry that our last programme together was the disastrous covering of the Giotto pass of Halley's Comet. As far as Apollo was concerned, I am sure that our early efforts were the best. There was one episode – during the Apollo 12 coverage – which might have been embarrassing. I was at one end of the studio, wearing a neck microphone, when the producer called me. 'Get across to the desk as quickly as you can – something interesting is coming up.'

I dashed across the studio, quite forgetting that my neck mike was in position – and the cord tightened up alarmingly between my legs. I said, instinctively: 'Oh, Ch***t, Mother, I've kn*****ed myself!' Mercifully, the microphone was not switched on...

Quarantining was strict for the first two Moon missions, but to be candid it was little more than window-dressing. Nobody believed that anything harmful could come from the Moon, but public opinion had to be taken into account, and people still remembered the first major science fiction television serial, *The Quatermass Experiment* (which with its two successors, *Quatermass II* and *Quatermass and the Pit*; remains unsurpassed, almost the whole of England watched it). But when Neil and Buzz landed, they were airlifted by helicopter to the waiting ship, and during this time they were exposed to the air, making the whole process rather pointless. Subsequently I was at Mission HQ when the first lunar rocks were examined, they were in a sealed box, and the initial examination was by robot; as expected, the lunar samples turned out to be very similar to Earth volcanic rocks. Quarantining was abandoned after Apollo 12, because the dangers from the Moon were by then regarded as nil, but it must be added that Mars will be a very different problem, because even though there are no little green men Mars may not be a totally sterile world.

(Did you see the film *Apollo 13*? It was well done, and kept pretty well to the facts. There was one unexpected actor, in a non-speaking rôle; the commander of the US ship which picked up the astronauts after splashdown was played by James Lovell, Apollo 13's commander. He protested that he wanted to be in his own film! It is sad that he never did get to the Moon, indeed, he never again went into space, but he had played a great part in the whole Apollo programme.)

The rest of the Apollo's were less fraught, but each one brought new breathtaking views; remember Apollo 15 in the foothills of the Lunar Apennines, and Apollo 16 in the rough southern uplands. I stayed in Television Centre until the last mission, No. 17, and this was partly my own doing. There was little I could do at the actual site, but if communications between Cape Canaveral and TV Centre had broken down for any reason I would have been able to 'hold the fort' until contact was re-established. This did once happen, and I had to talk for about ten minutes without really knowing what was going on, but I hope I sounded convincing.

When it became clear that Apollo 17 would be the final mission, Tam Fry, one of the senior producers (by now there were so many producers that I lost count) said that it was only fair for me to see one launch. So, instead of remaining in Television Centre, I went with the rest of the team to Cape Canaveral.

The whole base was a hive of activity, with not only the scientists but also visitors, plus almost every television and radio commentator on the planet. There was a slight feeling of regret that this great adventure was about to come to an end; the decision had already been made to scrap Apollo's 18 to 21, because they could not add a great deal to that had already been learned – and sooner or later something was bound to go badly wrong. One of the Apollo 17 Moon-walkers, Harrison (Jack) Schmitt, was a professional geologist who had been trained as an astronaut especially for the mission. All the previous crews had known a good deal about geology, but the decision to send a professional was sensible, and it certainly paid off.

I was fortunate in that Apollo 17 was the only launch that took place at night. It ought not to have been but during the final countdown there were several delays, and by the time that the rocket took off the Sun had long since set. We in the BBC team were some way back around four miles from the spacecraft, but almost nobody was closer than that. We watched during those last moments of the countdown, and then 'We have ignition!' Even from afar the light was blinding – the firework display to end all firework displays. Seconds later we were hit by what I can only call a wall of sound. It was absolutely deafening, and yet there was an eerie quality about it which I find hard to explain. Then, slowly at first, Apollo moved away from the launch pad; as it rose it gathered speed and we followed it until it was lost to view. The Last Men on the Moon (so far) were on their way.

While on the Moon, Schmitt and the mission commander, Eugene Cernan, made an unexpected discovery – 'orange soil – crazy!'. I was excited, because – like the astronauts – I thought that this might indicate recent volcanic activity. Disappointingly, the colour turned out to be due to large numbers of very ancient glassy 'beads'.

The final blast-off was dramatic. An automatic camera had been left on the Moon, and as Apollo 17 soared upward the gases from its exhaust showed up well on the television screen. It was the end of an era, but at that time I doubt if anyone believed that no more men would make the journey before the end of the twentieth century. There had been expectations of lunar bases, with huge domes rising from the lava plains, and veritable cities inhabited by scientists of all nations, north, south, east and west. This has not happened yet, but it may well become reality before many more decades have run their course, unless of course humanity decides to wipe itself out.

Nobody has made two trips to the Moon, and by now the surviving Apollo astronauts are well over the age limit. (Yes, I know that John Glenn went into space when in his late seventies, but that was into Earth orbit, and in any case John Glenn is no ordinary septuagenarian, believe me.) All the men of Apollo were American. Russia's rocket programme was in trouble, and after a disastrous explosion they gave up any idea of being first on the Moon; had their plans worked the first man on the Moon might well have been Alexei Leonov, who has all the right qualifications. I have a tremendous liking and admiration for Alexei, who is one of the most experienced of all spacemen. I have broadcast with him on quite a number of occasions, and joining him for vodka afterwards is always a joy!

After Apollo, the main emphasis shifted to the unmanned exploration of the planets, and I was involved in much of the television coverage. It was an exciting time, but not everything went well. A Mars probe, Mariner 3, was lost because someone forgot to feed a minus sign into a computer, which makes quite a difference. The Russians lost another Mars mission, Phobos 1, in 1989 because a technician sent it a wrong command; Phobos 1 promptly switched itself off in disgust and was never heard of again (I have often wondered whether the offending technician was quietly removed to a salt-mine in Siberia). The worst blunder of all was with Mars Climate Orbiter, a spacecraft launched in 1999 to improve our knowledge of the Martian surface. During the approach manoeuvre, instructions were sent to it in Metric units, blissfully regardless of the fact that it had been programmed to work in Imperial. The result was predictable: Farewell, Climate Orbiter. As I remember saying, this was yet another case of the evil of creeping Metrication!

Two planets were singled out for particular attention: Venus and Mars, initially in that order because it was often thought that Venus might be more promising as a potential colony. It is twenty miles closer to the Sun than we are, making it more outwardly attractive than the frigid Mars, it is about the same size as the Earth, and it has a dense atmosphere. True, it was known that the main constituent of the atmosphere was carbon dioxide, which is a 'greenhouse gas', but it did not seem outrageous to suggest breaking up the carbon dioxide and releasing free oxygen, which would cool the whole planet quite quickly. As for the surface – well, nobody knew. Sir Fred Hoyle believed in oceans of oil 'beyond the dreams, of any Texas oil king'; Fred Whipple and Don Menzel preferred oceans of water, which would be fouled by the atmospheric carbon dioxide and turned into soda-water (finding whisky to mix with it did not seem likely, but one could always hope).

Around 1960 I discussed all this with Don Menzel, who seemed optimistic. Shortly afterwards I gave a lecture about Venus during which I made a dozen definite statements, each of which was supported by the best available evidence and every one of which was subsequently found to be wrong. I suggested that since life on Earth began in our seas several thousands of millions of years ago, when the air was rich in carbon dioxide and lacking in oxygen, the same thing might even now be happening on Venus, so that in the fullness of time advanced life forms would evolve.

Not so! The Russians launched successful spacecraft, and managed to obtain television images which told a very different story. I remember seeing the first surface picture, which had been sent through to me direct from Moscow. The red, rocky landscape was hardly welcoming; the temperature is not far short of 1000 degrees Fahrenheit, and the atmospheric pressure is crushing – some of the earlier Soviet probes had been literally squashed as they dropped into the denser air on their way down. Not surprisingly, the main attention of the space-planners was at once turned away from Venus and directed back toward Mars.

Lowell's Martians, and his canals, are non-existent, but we cannot yet say definitely that Mars is completely sterile. The Viking probes of the 1970s had found no trace of life, but the jury is still out, and we will not get much further until we can bring back samples from Mars and analyse them properly. Of course, the 'face' was still discussed in the 1990s. It is true that a rock imaged from Viking orbit did look uncannily humanoid, and the flying saucer enthusiasts were in full cry; the Martians were there, or own Governments were busy on a monumental cover-up etc., etc. It was no use explaining that we were dealing with nothing more significant than light and shade effects, even when later images taken from a different angle removed any likeness to a face, humanoid or otherwise. One simply has to give up. Certainly Mars did once have surface water, and no doubt an atmosphere much denser than it is now, but whether palaeontologists will find any fossils remains to be seen. Meanwhile, I am confident that before long the UFOlogists will be able to link the 'face' with our own Altanteans, Lemurians and crop circles, as well as the conventional alien abductions.

Another 1990s topic was that of ice on the Moon. Some of the polar craters are deep, and their floors never receive any sunlight, so that they remain bitterly cold (they must be among the most desolate places in the Solar System). On 25 January 1995 NASA launched an unmanned spacecraft, Clementine (named after the character in the old song, who was 'lost and gone forever'), which entered lunar orbit and sent back excellent pictures. Instruments on board reported that the floors of some of the permanently shadowed craters were covered with ice, and the media became excited; ice could be turned into water for the use of future colonists... I was sceptical from the outset. First, Clementine had not detected ice directly, but only hydrogen – and this probably came from the 'solar wind', made up of particles being set out by the Sun all the time. Some reports suggested that water might be plentiful enough to fill large lakes, and there were no immediate denials from the authorities, because this sort of thing is helpful when applications for extra funds are being put in hand... Nobody seemed to know how to 'mine' the ice mixed in with rock. It would be a rather ticklish operation.

In any case, how could ice have got there? Brought by a comet? No; a crashing comet would produce enough heat to vaporize any ice. Sent up from inside the Moon? No; nothing could be less likely. However, NASA persisted, and sent up another unmanned probe, Prospector, on 6 January 1998. For over a year it orbited the Moon, and was then deliberately crashed on to the floor of a polar crater in the hope that the débris hurled up would show indications of water. It didn't. I may add that another immediate sceptic was Jack Schmitt, and as he has been there I feel that his opinions should be taken very seriously!

My visits to NASA became much less frequent after the end of the Voyager programme, but in 1990 I went back to Cape Canaveral – at NASA's invitation, I am proud to say – for the launch of the Hubble Space Telescope, a 94-inch reflector which orbits at nearly 300 miles above the Earth. Edwin Hubble, after whom the telescope is named, was one of the greatest of all twentieth-century American astronomers; it was he who showed that the objects once called spiral nebulae are in fact galaxies, millions of light-years away; and that the whole universe is expanding. I met Hubble several times around the late 1940s, though of course I did not know him well. He had the reputation for being rather austere and overbearing, but as a young amateur astronomer I found him very approachable, and what I saw of him I liked very much. He died rather suddenly in 1953.

The Space Telescope launch was dramatic; there were several

last-minute hitches, but at last the telescope soared aloft. I watched it together with several world-famous astronomers. One particularly eminent designer of scientific instruments, whom I refuse to name, decided to take some photographs. Unfortunately he had little idea of how to load a pocket camera, and I had to show him where to put the film in. However, all was well in the end.

It was only when the telescope was safely in orbit that an appalling discovery was made. The mirror – the essential heart of the telescope – was faulty; it had been wrongly figured, and the pictures sent back were blurred. It was then, I feel, that many science writers behaved very badly indeed. Cheap jibes were rife: 'Flop of the Century' and 'Hubble Needs Spectacles' were typical headlines. I can only say that in *The Sky at Night* programmes we never descended to that level, and we went out of our way to stress that even though the telescope was not as good as had been hoped, it was still a major advance. I know that the HST team appreciated this, and in the end we had the last laugh. A repair mission put the telescope right, and today it is performing even better than was originally expected.

I wonder why some writers, and many interviewers, are always anxious to publicize the failures rather than the successes? In the media, the Hubble Telescope received much more attention when the fault was discovered than it did at the time of the launch. Sadly, this is typical of modern attitudes.

When I was born, the world's largest telescope was the Mount Wilson 100-inch reflector; radio was becoming popular, but there was no television and no electronics, while air travel was very much of a luxury. Life has changed beyond all recognition between 1923 and 2003, but whether it will show an equal change between 2003 and 2100 is less certain. I repeat that it all depends, ultimately, upon world leaders. In many ways I am sad that I will not personally see very far into the twenty-first century, but about this I can do nothing! Mind you, I did know one man (actually an astronomer) who was firmly convinced that he was immortal. When he died, at the age of eighty-six, he must have had a nasty shock.

One thing which may make a vital difference to the new century is the exploration of Mars. Chilly though it may be, and lacking in breathable air, it is still not too hostile, and it must be our first target once we have really established ourselves on the Moon. At least Mars – unlike the Moon! – has plenty of ice, and to the best of my knowledge there is nothing dangerous there. The weak link may be the human body, and there are always fears about undesirable radiations from the Sun and from space. Frankly, the only course here is to take what precautions we can, and hope for the best. Obviously the first Mars travellers will take the greatest risks, because they will not really know what to expect – and there is a slight similarity here to Yuri Gagarin's initial space-flight in Vostok 1, getting on for half a century ago now.

'Terraforming' Mars and turning it into a sort of second Earth is a delightful idea, but whether it can ever be achieved is debatable, and it is wildly beyond our capability at the present time. So a Martian colony will have to be limited, and there can be no chance of using it to solve Earth's menacing over-population problem.

If a Martian colony is set up, I wonder how it will develop? It means that those who go there will have to be unselfish, able and mentally stable, characteristics which are seldom found together in the same person. Yet it might lead on to a general change in outlook, and it is even possible that this might spread to Earth. Bear in mind, too, that we cannot be sure that a boy or girl born and reared on Mars will ever to able to come to Earth, where the gravity is so much stronger. How would you feel if, suddenly, your weight increased by a factor of three? The 'Martians' may have to resign themselves to staying either at 'home', or visiting the Moon.

Time will tell. Meanwhile, I have an intriguing suggestion. Our forbears were in the habit of exiling undesirables to Australia, so why not use Mars in the same way? Round up all the world's nastiest people – Saddam Hussain, Osama bin Laden, Robert

Mugabe and the rest – put them in a spacecraft, deport them to Mars, and let them fight it out. The results would be interesting, and on Earth, to quote W. S Gilbert, 'they never would be missed.'

Well, it's a thought!

25 Coasting Along

During the time when probes to Venus and Mars were very much in the news, there was one important event in my life; my mother's 90th birthday, on 27 June 1976. It was a joyous occasion, and although she was rather frail she was in great form; she was able to move around easily, and mentally she was 100 per cent.

We had always been exceptionally close, partly because we had the same sort of outlook but also, I suspect, because I knew I would never marry. Herr Hitler had seen to that; there was nothing to be done about it, and Mother understood only too well. When she was well into her eighties she looked absurdly young, and I reckon she could have passed for mid-sixties. I remember one episode when she was 87. I had been in London, and came home around five o'clock; Mother was in the front garden, and the conversation went like this. She began it: 'It's been dull today – let's have a party,' 'What a good idea!' 'Right, you see to the drinks, and I'll get on to the phone.' By half-past seven there were at least thirty people in the house.

For her 90th we threw a party at the Selsey Hotel, run by a friend of ours, George Castens, who had been a distinguished submariner during the war. It was quite a gathering; surviving relatives (mainly cousins), old and new friends, and many locals. It had been a blazing hot day, and the Sun shone down from an azure sky. Not a day to be forgotten…

She was not trained as an artist, but her watercolours were

good as well as unusual. Some were serious scenes, but she was at her best with the creatures she called 'bogeys', all of which were pleasant rather than scary. She was doing them when she was a girl, and when I came along and took an interest in astronomy I persuaded her to go back to painting. She produced amazingly individual pictures; a baffled Moon man looking down at a crashed Sputnik, a warden directing traffic in the crowded asteroid belt, a joy-rider on a comet, and motor cyclists riding round the rings of Saturn (that was her own favourite). All the paintings were framed, and were displayed in Farthings; every year we had a print made from the latest painting, and these prints were sent out as Christmas cards. This went on until 1991; when she had to give up painting, I abandoned Christmas cards. I just hadn't the heart to send any others.

However, in 1972 a publisher was having dinner with us, saw the pictures and asked if they had been published. They hadn't; so we collected them, I wrote a text round them, and *Mrs. Moore In Space* came out. It sold well, and at the moment (January 2003) is being reprinted. I wish I had inherited this talent; if I had one scrap of artistic ability I would try to develop it, but I am so hopeless that there isn't really any point. I can do line drawings, but so far as I am concerned perspective is a closed book.

We were lucky in one respect, as I have said; Mrs. Woodward ('Woody') came to us, initially as a housekeeper, but quickly as a very dear friend. Life went on, but of course, Mother was less mobile than before, and I stayed at home, cutting out all foreign trips apart from the brief forays to NASA for the Voyager passes of Jupiter and Saturn. Normally I would have gone to the total solar eclipse of 1980 because the track crossed the island of Mombasa, off the coast of East Africa, where my parents had lived between 1919 and 1922; apparently my mother was long remembered as having been responsible for the planting of a row of flamboyant trees along the side of what was then the main road. Mombasa has since changed from a Portuguese village into a bustling modern city, and I do not know if those flamboyant trees are still there; I hope they are.

Only toward the very end did Mother begin to fail physically; mentally she never did. Near the very end Woody and I desperately needed help. I called Peter Cattermole and Iain Nicholson, and they dropped everything – they were with us in a matter of hours; that is the sort of friends they are, and they saw us through. Mother died peacefully, at home, on 7 January 1981; needless to say I was with her. Nothing could ever be the same again. The only consolation is that she had a long innings.

Woody's death, a few years later, was a total shock, she was ill for less than twenty-four hours. Her son Barry, a successful business man, came to live in Selsey with his wife Jennifer and their two boys, Neil and James, and it was great to have them close by, but there came a time when they decided to move to Yorkshire, and I said a sad goodbye.

On 4 March, 1993, I passed from my sixties into my seventies. This did not appeal to me in the least, because there is absolutely nothing to be said in favour of growing old; there ought to be a law against it. Actually I didn't start to feel ancient until my wretched frame began to give trouble, when I was 77, but obviously there were some things which had to be modified or abandoned. Well into my seventies I was playing cricket, and getting a good tally of wickets, but eventually my lame knee put paid to that, and I retired to the pavilion to act as scorer. (I don't like umpiring; if there is an appeal for 'caught wicket' I am never sure whether the batsman has touched it or not). In fact I had played much less in the latter years, because I would always stand down in favour of a youngster waiting for a game.

With Woody gone I was living alone, which I hated. Of course I had always known that this was likely to happen eventually, but it wasn't to my liking at all. At least I had many friends; Selsey is a friendly place. I also had Bonnie, who had four legs, was black and white, and said 'mew'.

I have always been a cat person. During my early years I had two, Ginger (guess what colour he was!) and Ptolemy who was as black as night. Ptolemy, who was my constant shadow, died of cat 'flu' when I was ten, and that was without any doubt the worst day of my boyhood; I just couldn't take it. Ginger was joined by another marmalade cat, Rufus, who was the kitten of a feral cat and was born under our garden shed. There were in fact four kittens; we found good homes for three, but Rufus made it clear that he was never going anywhere, and he had a blissfully happy life extending over twenty years. Later came Smudgie, Beno and then Bonnie, who was a real character. She insisted on sleeping on my fax machine, and I remember an irritated call from New Zealand: 'I'm trying to send you a fax. Will you kindly tell your cat to get off it?'

A catless house is a soulless house. The only problem is that cats have such short lives, and it is heartbreaking to have to say 'goodbye'. When Bonnie came to the end of her days, in her twentieth year, I swore that I couldn't go through that grief again, but my godson Adam, who had come to make his home with me, had other ideas; he is as silly about cats as I am. In the autumn of 2000, when I was in hospital having a knee operation (which didn't work), Adam – then twenty, and in the middle of his first degree in computer science, at the University of Sussex – went to see some kittens which had been advertised for sale. One small black and white kitten, a few weeks old, was named Jeannie. Adam looked at Jeannie. Jeannie looked at Adam. It was love at first sight. Fifteen pounds changed hands; one kitten changed hands …

I can honestly say that Jeannie is the most adorable little thing I have ever come across. She looks lovely, and has a nature to match; I couldn't bear to be without her. She will outlive me by many years, but very careful provision has been made for her when I am no longer around. She even has her own large enclosed garden complete with fountain. The notice in our hall says it all. The house is maintained entirely for the convenience of the cat!

When I say that Adam is my godson, that isn't strictly true; he is my godson's son. In my teenage years, and after the war, I had two very close friends, Ian Corrie and Pat Clarke; we were of the same age, and we saw a great deal of each other. Tragically, both died of cancer when they were around sixty. (What a curse cancer is – and why do not our politicians provide enough money for research, instead of spending it on bombs?) It is ironical that I am still here, while they are not, because physically I was never the equal of either of them.

Pat's two sons, Lawrence and Matthew, have always played a great part in my life, and still do; I find it hard to realize that we are not actually related. Ian's son, Nick was my godson. Tragically, he died young; his wife was Irish, and went back there, so that Adam elected to join me. About this, I couldn't possibly be happier.

Another arrival at Farthings was Roger Prout, who had been a great friend for around forty years. He had been living in Chichester, but had to move house, so I suggested that he might move in – and he did; luckily Farthings was the right sort of house, and my second study became Roger's. By profession he has been a lawyer, working in a senior capacity with the Chichester Council. I say 'has been', because he has retired to concentrate upon what he really loves – opera. He was for a while with the Welsh National Opera Company, and what he doesn't know about this form of art is not worth knowing. Also in Selsey, a few doors away, live Jim and Catherine Galloway; Jim is a retired mathematician, while Catherine had made her name as an operatic soprano before abandoning it to give all her time to her husband and children. Now that her children have fled the nest, and Jim is happily retired, Catherine has re-started her career, with great success; also, she has teamed up with Roger to present concerts – Catherine to do the singing and Roger to read suitable poetry. It shows signs of working very well.

I am no opera buff (I don't count Gilbert and Sullivan). With

Roger, I went to a performance of *Dr. Broucek Goes to the Moon*, an opera by the Czech composer Janáček. I still do not know whether it was sung in English or Serbo-Croat, but I am sure that it would have made no sense in either language, while the musical score sounded to me like a village orchestra tuning up. I think I will leave it at that.

But if opera leaves me cold, acting does not, and I have taken part in a great many shows – as an amateur, of course, and I am no good in anything except farce. At East Grinstead, and subsequently at Selsey, I became involved in the Christmas pantomimes, and I specialized in demons. Once, in my East Grinstead days, I caused rather a stir. We had been asked to take our pantomime to Tunbridge Wells for a special performance in aid of the local hospice, and we duly did so. Of all those taking part, I was the only one to use old-fashioned grease-paint, and I had given myself a green make-up, plus horns and a tail. When we were ready to go, we found that both our minibuses had broken down; no taxis were available, and all we could do was to catch the ordinary bus. There was no time to change, and so I had to make the journey in full make-up. One old lady prepared to board the bus when we stopped at Hartfield; she saw me, screamed and fled. I suppose it must have been a little disturbing. At least I looked realistic; demons suit me.

Roger is also a very fine drama producer. In 2002, at Funtington (near Chichester) he put on five short one-act plays that I had written; we called the show *Quintet*, and it had a very strong cast, far above the usual amateur standard (several members had had professional acting experience, but had given up the stage to become medical doctors). The plays were all different; for instance *Total War* set in World War Three (fortunately the two opposing commanders of forces on a strategically important Pacific island have been at school together); *Quangoland*, a Civil Service play about a Government department which has existed for thirty years doing nothing except administer itself); *Shades of Tolstoy*, a Russian drama in which all the characters shoot themselves one by one, leaving only the hapless Vladimir Nastikoff, who has to be shot by the compère in order to end the play... you get the general idea. I am glad to say that the little theatre was packed out every night!

I did once act with professionals. Near Guildford, a repertory company was putting on a farce by L. du Garde Peach; the man playing the German general was taken ill; so was the understudy, and I had a frantic phone call asking me to come to the rescue. I was on only in one scene; I had a day to learn my lines, and I went on. I was then asked to play for the rest of the fortnight's run. I enjoyed it, though I am well aware of the wide gulf between the experienced amateur and even the minor professional.

Over the years I have met many 'showbiz' people, and I have been on radio and television with some of them. I went on TV with Morecambe and Wise – what fun that was. I was one of a line-up of people who were fairly easy to recognize, and we danced solemnly around, dressed in dinner-suits; that recording is still shown occasionally even now. Magnus Pyke – remember him? Twice we joined forces as a couple of cross-talk comedians, and on the second occasion we 'tip-toed through the tulips', billing ourselves as the Hit Parade of 1988. Magnus, eminent scientist though he was, was never afraid to laugh at himself, and I think I can say that the same is true of me. Taking oneself too seriously is a great mistake, and this is where so many politicians fall down.

Some performers stand out in my memory. I knew Arthur Askey well – one of the most delightful people I have ever met, and the first of the great radio comics. *Band Wagon* became a cult, and Dickie Murdoch was an ideal foil to Arthur. Kenneth Horne was another splendid character; I broadcast with him on *Round the Horne* a day or two before his sudden death – he was much missed. Offhand I can't think of any current comedian to match Arthur, or Dickie, or Kenneth. I belong to an earlier generation and I may well be prejudiced, but note that the old-timers had no need to go 'blue', they could manage very well without that, and were clever enough to do so.

Will Hay, the comic schoolmaster of the pre-war films, was a very serious astronomer who discovered the white spot on Saturn in 1933; at one time we served together on the Council of the British Astronomical Association. I knew Richard Hearne for a very different reason. As Mr. Pastry he was a brilliant comedian, but his real aim in life was to help others, and he was untiring in his efforts. He founded the Mr. Pastry Fund, which did one thing and one thing only: build swimming pools for children suffering from polio – because swimming is something which can really help a polio victim as well as giving pleasure. I was co-opted at an early stage, and I am glad to say that we raised enough money to build a number of pools which are still in use. Not one penny was spent in administration. I remember once giving a lecture in aid of the Fund and arriving at Richard's house, very late in the evening, with two large sacks full of money! Yes, Mr. Pastry will be long remembered, and he is one of those rare people who can never have had an enemy. I wish I could have carried on the Fund when he died, but I simply haven't got that sort of organizational ability. I can be a back-up man, but not a leader.

I recall having some drinks in a BBC bar with five young men who were just starting to become known. Their music was not my music, but I liked them, and mentally classed them as decent, talented lads who would go a long way. Well yes – the Beatles certainly did that. I am sorry that I never met them again.

My orbit crossed that of Michael Bentine in 1940, and for various reasons we saw a good deal of each other during the next few years; when the world returned to normal (or as normal as it ever is), our friendship remained exceptionally close. His life was touched by tragedy, but he was always ready to meet any challenge. The Goons are part of history; Michael was not with them all through their career, but he was the cleverest of them all, and some of those early recordings are as hilarious now as they were when they were first made. Michael once launched the BBC Television Centre into space, with Richard Dimbleby on board and me carrying out a running commentary. Subsequently the BBC sent him an official rebuke: 'Television Centre is not to be used for the purposes of entertainment.' I can assure you that this letter really exists; I have seen it!

In a *Sky at Night* studio (I think it was Studio 7) we once tried to demonstrate zero gravity, and we were flown around in specially designed pieces of apparatus. I have a photograph showing us wallowing around, dressed in cumbersome space suits and looking a little like aerial sea lions. During another programme, filmed in my Selsey study, we were discussing alien life when an alien appeared at the window, asking us whether we really did represent terrestrial life. I think it worked rather well.

Another experiment was not so successful; it was an investigation into telepathy. Michael and I were both interested in that sort of thing, and of course we were very much on the same wavelength, so that we were regarded as good subjects. A TV camera was trained on Michael, in London, and another on me, in Selsey; the idea was that we would try to transmit pictures to each other telepathically. I have to admit that it was a total failure. For example, Michael tried to send me a picture of a flying bird, all I managed, after intense concentration, was a pattern which bore a slight resemblance to a demented spider. At least we tried.

Michael died of cancer; I was with him two days before he went. One of the last to see him was Prince Charles; only a few months earlier Michael and I had been the Prince's dinner guests, together with Eric Laithwaite (now also dead) and Richard Gregory. Later, after Michael's death, I was called to the Palace to be dubbed a Knight. Prince Charles took the ceremony; I knelt to receive the tap on the shoulder; we then talked for the conventional thirty seconds – about Michael. As the only Peruvian-Scottish Old Etonian, Michael Bentine was a 'one-off'. Remember too that he was a first-class engineer, as he showed when testing a hovercraft down the Nile – and at one stage in his career he saved a great many lives.

Today, I have another great friend who is so famous in his own field that many people do not appreciate that he is a highly-qualified astronomer: Brian May. As the leader of the Queen group, he is widely regarded as the world's most gifted guitarist, but he had a first-class degree in astrophysics, and he retains his interest. When he first joined me in a *Sky at Night* programme, purely as an astronomer, many of his countless admirers were taken aback. He is a specialist in 'cosmic dust', and we are now planning a programme about the Zodiacal Light, the ethereal cone of radiance stretching up from the horizon after sunset, caused by sunlight being reflected from grains of material spread along the main plane of the Solar System. I would say that Brian, like Michael Bentine, is one of those rare folk with many friends and absolutely no enemies. We were all delighted when, in 2003, he was made an Honorary Doctor of Science. Congratulations, Dr Brian May.

Members of Parliament are not noted for their knowledge of astronomical matters – recall the UFO debate in the Lords, initiated by Lord Clancarty, which showed that W. S. Gilbert, in *Iolanthe*, was very close to the mark (even if not many of the Lords agreed with the Earl of Clancarty that flying saucers have a mooring base deep inside the Earth, and pop out through the Hole in the Pole). But there are exceptions, and one of these is Lembit Öpik, who sits as a Liberal (or Liberal Democrat, or whatever the Party calls itself this year). His grandfather, Ernst, was an Estonian who spent much of his career at the Armagh Observatory, and whom I knew well and liked immensely, even if we did not always agree in matters of science (he was convinced that sunspots were associated with political revolutions, and that the Russians had never sent any men into space). Lembit is not officially an astronomer, but he knows a great deal about it, and is concerned with the threat to Earth by asteroid bombardment.

Wandering bodies are more common than used to be thought, and certainly the Earth is vulnerable; it is widely believed that a massive impact 65 million years ago changed the climate completely, wiping out the dinosaurs.

There are periodic scares, as in July 2002, when a report claimed that a mile-wide asteroid, 2002 NP, would hit us in February 2019 and cause global devastation. During the days of 24 and 25 July I made twenty-six broadcasts about it, on national and local radio (I counted them). Later calculations showed that there will be no hit, but what would we do if we saw a massive impactor on a collision course? Try to break it up, try to divert it by exploding a nuclear bomb close by it, or simply sit back and pour another gin and tonic? Lembit is trying to tell Government leaders that there really is a risk, and of course he is right. Like good Boy Scouts, 'Be prepared'.

Sporting personalities? Yes, of course. Football is a closed book to me (both soccer and rugger), but as a Lord's Taverner and Life Member of Sussex CC I have come across many cricketers. I have yet to encounter a first-class cricketer whom I have disliked. The Taverners are great fun, as well as raising large sums of money for cricket grounds and young players. At one function I had a long conversation with Don Bradman, who was forthright, friendly and above all modest; I was always sorry that he did not make four runs in his last Test innings, to give him a final average of 100 instead of 99.96. My favourite Bradman story was told to me by a radio interviewer, who asked: 'If you were playing against England today, on different wickets and with slightly different rules, what do you reckon your average would be?'

Bradman thought for a moment. 'Against the present England team, I don't think I would average more than about 55.' He paused. 'Mind you, I am eighty-seven!'

I can't be sure that the story is true, but I think it is. It sounds right. At another function I was asked to compile the best team from players I had actually seen, which in my case cuts out

Hobbs, Sutcliffe, Rhodes and of course W.G. I came up with the following list: Boycott, Atherton (capt), Botham, Compton, Randall, Greig, Ames (wkt), Tyson, Statham, Underwood, Wright. For No. 1, I hesitated between Boycott and Hutton, and to include Laker would mean omitting Greig. I am sure that many people will have different ideas – but after all, I'm only a village-greener.

There are Taverners of all kinds. I think most people know that Brian Rix is a first-class cricketer, but not everyone realizes that, for instance, Clement Freud is a batsman of County standard, and Jasper Carrott an excellent fast bowler. I am only sad that my own playing days are over.

Tennis was another love of mine. On a Club court near my home I once tried to take Jaroslav Drobny's service. I could see this racket come down, and heard the net behind me shaking, but what happened in between I never found out. I also recall a Wimbledon Players' function, held some years ago before the Championships, in aid of Cystic Fibrosis. The player who stayed all the time, doing all she could to help, was Billie-Jean King.

Then snooker. I was taken to London Airport to face Steve Davis, the idea being to collect cash for handicapped children. He broke off, and left me a very easy pot into a middle pocket. I sank it, and of course followed it in; Steve commented that that required great concentration. He then made a break of about fifty, and left me another easy pot, this time I sank not only the red, but also the blue. In desperation I called for the cue ball to be cleaned, which seemed to amuse the audience. The eventual score was Moore 0, Davis 133 – but at least we made well over £500 for the children.

Also in aid of children's charities, I went to Twickenham Rugby Ground, at the behest of Norris McWhirter, to play 200 simultaneous games of draughts; the boards were all laid out ready and trekked among them. I am a chess player, but not really a draughts player; however, I won more games than I lost. On the debit side, I was totally defeated by a boy of six. Afterwards I challenged him to a revenge match – and again he won. I had to admit that he was much too good for me.

Norris, as most people know, is a key figure in the Freedom Association, and is one of the relatively few well-known figures prepared to 'stand up and be counted'. In fact, he speaks for Britain, and it is a great pity that some other prominent folk lack the courage to follow his example.

Quite apart from *The Sky at Night*, I took part in various television and radio programmes, not all of which were serious. As I have had cause to mention, Devon and Cornwall County Councils did not easily forgive me for announcing a postponement of the 1999 total solar eclipse. On another April the First I went on Radio 4's *Today* programme (then run most entertainingly by Jack de Manio, and very different from the po-faced political programme we have now) and said that at 10.00am Venus would pass behind Jupiter, and this would cause a slight loss of weight, which you could check by jumping up and down. Amazingly, the BBC phones were jammed for hours, and we had some very interesting calls, notably from courting couples. But this was before a similar broadcast by my old friend Professor Samuel Tolænsky, FRS, who fried onions in front of a TV camera and broadcast smells. To me, this equals Richard Dimbleby's immortal spaghetti farm which bamboozled many of his devoted *Panorama* viewers when he gently pulled their legs on another April Fool's Day.!

For some time I was ostracized from BBC's *Any Questions*. A very Politically Correct woman MP told me, icily, that immigrants coming here were 'as British as you and me'. She wasn't ecstatic when I pointed out that if a cat has kittens in a pigsty, they don't turn into piglets. I was forgiven eventually. I am afraid I also raised eyebrows later, in June 2002, in a TV programme called *Room 101*; you have a large dustbin, where you are asked to dump anything or anybody that you particularly feel should be removed from the scene of operations. My first candidate was Mr. Carey, then Archbishop of Canterbury, for his abject refusal to condemn foxhunting. (He resigned shortly afterwards. It is perhaps a little unlikely that his departure had anything to do with *Room 101*, but at any rate he was no loss!)

I also disposed of impenetrable wrapping paper and maybugs, those huge flying cockroaches. Yes, I know they are harmless, but I have a deep-rooted terror of them, and am convinced that they will get me in the end. But I also nominated all female radio news-readers; when you are driving a car, and want to hear what is happening in the latest wars, you can't cope with their piping little voices. I had quite a stack of letters about that one!

I had plenty to do during the 1990's. *The Sky at Night* continued as of yore; I continued observing – my telescopes were in use every clear night; I was a regular attendant at meetings of the Royal Astronomical Society and the British Astronomical Association, of which I was Director of the Lunar Section. I had many visitors to my observatory, notably young enthusiasts, and I would never turn anybody away. I was involved in local activities as well as cricket and tennis, and one of these was the Birdman Rally, which had a bright start, a lively career and a sad end.

The idea came from George Abel, formerly of the RAF and then running a photographic shop in Selsey. The plan was to challenge people to jump off the Lifeboat Pier (the only pier available) and fly a hundred yards with the aid of artificial wings. He co-opted helpers, including me, and we put out a preliminary notice. There was immediate interest, and we hired the RNLI pier for an afternoon in August. We chose a time when the tide was suitable, and the number of entrants was surprisingly large – some serious, some not. Hot air was not allowed, and neither was any kind of mechanical aid, while ornithopters (wing-flappers) had to be worked by muscle-power alone).

Radio, television and the Press turned up; we organized marshals, and sent helpers round with boxes to collect cash for the RAF Association and the local hospice. We also had trained rescuers standing (and swimming) around in case anyone got into difficulties, though actually this never happened, because all the leapers were thoroughly at home in the water. We collected a decent sum, at once handed over to the hospice and the Selsey branch of the RAF Association.

After that the Rally was held every August, and became well known. Some of the tries were hilarious, for instance there were several pedal-driven helicopters, and kite-like contrivances which were quite impossible to control even before being launched. Above all, I remember a retired Wing Commander, who came over specially from his home in the West Country, bringing a beautifully made and elaborate flying machine of the helicopter variety. It took him at least an hour to assemble, after which he took it up to the take-off point on the pier and waited for the wind to be 'just right' (as he was at pains to point out, a few knots' difference in windspeed would be important). Everyone watched, mesmerised. Eventually he leaped – and the entire machine came to bits; the Wing Commander's descent into the sea was to all intents vertical. Needless to say, he was greeted with a deafening round of applause. 'Well,' he said philosophically, 'better luck next time!'.

Nobody managed the required distance, but the Rally gained momentum, and we went from strength to strength. The Japanese told us that they were sending a team all the way from Tokyo, confident of success. Then the blow fell. The RNLI officials told us that we could no longer use the Lifeboat Pier, even though we wanted it for only one afternoon in August and were ready to pay a fee. We pleaded with them, and two of us even drove down to the RNLI headquarters in the West, but to no avail. We looked for another pier, but there wasn't one. So we had to admit defeat, and the Birdmen of Selsey became the Birdman of Bognor. The Rally is still held every year, and is very popular, but somehow I have never felt inclined to go. Incidentally, the loss of funds from the Rally led to the demise of the Selsey branch of the RAF Association.

A Biblical society once pressed me to attend a meeting in Selsey, mainly to show that Darwinian evolution is rubbish, Creationism is true, and that this can be proved by studies of the Bible. I did go, but I kept a low profile until I was asked to speak. I fear my address did not go down too well, because I pointed out that the Bible can be interpreted to mean almost anything, and that by using selected quotes from the New Testament I could give conclusive proof that Jesus Christ was a scratch golfer. He was adept at finding lost golf balls: 'Ask, and it shall be given you; seek, and ye shall find' (Matthew 7, 7), but he did occasionally fluff a drive ('And some fell on stony ground, where it had not much earth' (Mark 4, 5.) He must have managed the occasional hole in one ('Rejoice yet in that day, and leap for joy' (Luke 6, 23), but has compassion for the poor wretch who takes thirteen at a par four hole ('Wherefore comfort one and thee with these words') (Thessalonians, 5, 18). If your opponent has the temerity to ask if he may borrow a club, 'Give it to him that asketh thee' (Matthew 5, 42). Finally, he was not a slow player: 'Look behind, I come quickly' (Revelations 3, 11). Straightforward enough, but I didn't feel that the audience was entirely on my wavelength, and I wasn't invited to their next gathering.

In 1998 my old friend Robin Vallier, who is actually a concert pianist but who organizes 'events', asked me to undertake a lecture tour round Britain. I had never done this sort of thing before, mainly because of lack of time, but it sounded rather fun. We also presented what we termed 'space concerts', with a small orchestra playing music I had composed – my only active part was in playing the xylophone. The tours were well received (at least, no eggs were thrown) and I made two more speaking tours in 1999 and 2000. I had to miss 2001, because I was not fit enough, but now (2003) I am doing what may well be my final effort. Mind you, I did once go to give a lecture to be greeted by an audience of 0. I had gone to the right place, at the right time, to give the right sort of lecture to the right people. Unfortunately it was the wrong year. I was 365 days too early... These things happen. However, I did eventually turn up on schedule, and at least it was better to be a year early than a year late.

26 Indoor Stars

When I was young, in the earlier part of the twentieth century, there were three events which seemed so far ahead that they might just as well have been in the infinite future: the return of Halley's Comet (1986), the English total eclipse of the Sun (1999) and the next transit of Venus (2004). The first two have happened, I saw the first, and would have seen the second but for inconvenient clouds over Cornwall at the vital moment. Event No. 3 is pending, though whether I will still be around then remains to be seen.

What I could not know in my youth was that I would host the longest-running television programme of its kind anywhere, and that in April 2002 this programme would have its 45th birthday. The edition of December 2002 was our 600th, a record which will not be beaten for a long time, if it ever is. (Yes, I know that there are some programmes, such as *Panorama*, which began before we did, but they have had different presenters, and few of them have had completely unbroken runs.) I have not missed a single programme, and I know that in this respect I have been lucky. I have avoided being ill at the wrong moment, and I have never run foul of any BBC strikes.

Near the turn of the century three things happened to me, all of which were completely unexpected. On 21 November 2000 I received a letter from the Secretary for Appointments at No. 10 Downing Street: 'The Prime Minister has asked me to inform you, in strict confidence, that he has it in mind, on the occasion of the forthcoming list of New Year Honours, to submit your name to The Queen with a recommendation that Her Majesty may be graciously pleased to approve that the honour of Knighthood may be conferred on you.' I was taken aback. Harold Wilson had given me the OBE and Margaret Thatcher had promoted me to CBE, but to be made a Knight of the Realm was something that had never crossed my mind. After all, what had I actually done? My only real research had been in connection with mapping the Moon, and this was more than forty years ago, quite apart from the fact that I was a very minor member of the lunar team. Since then I had been concentrating mainly on 'spreading the word' and trying to encourage others, and although I hope I had succeeded to some extent there were other people who had accomplished much more than I had.

I was shattered at the number of congratulatory messages I received when the New Year Honours were announced. I had over a thousand letters, not only from friends and acquaintances but also from many others outside my field, including politicians – a charming letter from Tony Blair, and also from William Hague, then Leader of the Conservative Party, and Sir Edward Heath, the former Premier. For the first and no doubt last time in my life I opened a special book, and in it fixed all the main letters. I felt overwhelmed, and even now I am barely used to being officially called 'Sir Patrick', I have mentioned the ceremony itself; 'my' boys, Adam and Chris, were there with me.

The second event was an equal shock: I was elected an honorary Fellow of the Royal Society, which is without doubt the most prestigious scientific society in the entire world. It was founded during the reign of Charles II, and its past Fellows have included Newton, Halley and other giants. Normally, Fellowship is given only to our scientific leaders, with the highest qualifications, whereas I am an amateur with no official degree at all (though I must add that the trouble there was due to Hitler, not me). When I was notified, I actually phoned the Society to say that there must have been some mistake – but there wasn't, and Sir Robert May, the President, confirmed that I was now fully entitled to put FRS after my name. I do so with the greatest pride.

No. 3 was the BAFTA Award, which again came as a major shock. The dinner at which the Awards were given was a marvellous occasion; on my table, with Chris, Adam and Roger, were several 'greats' of television, notably Esther Rantzen, for whom I have great liking and respect. It was an added honour when, as I have described earlier, Buzz Aldrin flew over from America to present me with the Award.

A knighthood, Fellowship of the Royal Society, and the BAFTA Award, all in a period of two years, – it still seems incredible, but the decisions were not mine.

In the sky we had two bright comets, Hyakutake in 1996 and Hale-Bopp in 1997. The second of these was magnificent, and hung in the sky for months, causing an immense amount of interest. I lost count of the number of people who came to see it from my observatory; it was prominent with the naked eye, but the telescope brought out intricate detail in the nucleus. If you missed it, look out for it again when it comes back in 2400 years' time.

Tom Bopp, one of the discoverers of the comet, came to England on a lecture tour, and stayed with me for several nights. It was a pleasant visit, and I well remember persuading him to try a glass of my home-made wine. Wine-making remains a mild hobby of mine, and you can use almost anything, within reason. Rose-petal is my favourite, but I also enjoy making pea wine, chiefly for the fun of offering it to people. 'Will you have a glass of…'

Finally, the last big project in which I will be involved: the South Downs Planetarium.

I am no stranger to planetaria. As I have said, I turned down an invitation to become first Director of the London Planetarium but in 1965 I did take one at Armagh, and ran the Planetarium for three years. It is fair to say that I left the Planetarium in a flourishing state, and it was one of the most popular tourist attractions in the entire Province. This continued for a long time, but then things started to go wrong. I do not know the full details, because frankly I rather distanced myself, but there were several problems. Also, the decision was made to abandon the Japanese projector, and replace it with a Digistar. A Digistar works on broadly the same principle, and it is more 'modern' so that it will do tricks such as putting on laser shows, but the star images are not crisp, and there is no colour. To me, a Digistar is not a patch on an 'old-fashioned' projector, but it all depends on the type of show which is to be given. London also changed to a Digistar, and the displays became more 'spectacular', though considerably less scientific.

Meanwhile, John Mason had a luminous idea. Why not set up a major Planetarium in Chichester? Not counting London, there was no full-scale Planetarium in South England, and it would be immensely useful for education as well as entertainment. The idea sounded ambitious, as it would mean starting from scratch, but initially it seemed reasonable to expect help from the Lottery and the Department of Education (of course, we ought to have known better). Everything hinges upon acquiring a good projector, and the cost was bound to be astronomical in every sense of the word. John Mason enlisted the aid of John Green, Roger Prout and Peter Fray; these four have done all the work, with me on the sidelines. We had a vast amount of enthusiasm, but no cash.

There was one hope. Armagh had mothballed its fine Japanese projector; could we lay our hands on it? I had the entré here, and there was no harm in trying. We had nothing to lose, because buying a brand-new projector was out of the question, and if we were unable to obtain the Armagh instrument it was painfully clear that our Chichester Planetarium would be stillborn.

Negotiations were opened. Let me say at once that the people at Armagh could not have been friendlier or more helpful, and as a start they set a price on the projector far below what it was actually worth. We paid a deposit, but when time ran out we did not have the shortfall, and Armagh would have been fully justified in keeping our deposit and abandoning the whole deal. They did nothing of the kind; instead they waited, and waited and waited – until at last we had mustered the amount we needed. Armagh then handed the projector over, and it was cautiously shipped across to Chichester.

Very sadly, the Armagh Planetarium is now out of action, and the Digistar has been dismantled and taken out of the dome. This is unfortunate, to put it mildly; it should never have happened, and from all accounts the technical organization and management after about 1990 was not good. I hope that before long the situation will change, but so far it has to be admitted that the future does not look particularly bright. At least, the Japanese projector is in full use, and our Planetarium – named, at my suggestion, the South Downs Planetarium – owes its very existence to Armagh.

We made inquiries in Chichester, and found that both the local Council and the Director of Education backed us to the hilt. So did the local paper, the *Chichester Observer*, whose Editor, Keith Newbery, has been immensely supportive throughout; needless to say, this has been of the utmost value to us. We acquired a building just outside the grounds of the Chichester Boys' School, and were told that we could share the school car park until we could obtain one of our own. Local industry chipped in, and we were given expert help in many ways. It was, incidentally, lucky that the four key members of the team had varied skills; Roger was a lawyer and Peter a businessman. John Green had just retired from a senior position on the CID and devoted virtually his whole time to the Planetarium, while John Mason, as a highly qualified scientist, looked after all the technical side. I was the only one who felt rather useless, so I concentrated on fund-raising.

There was a great deal that could be done. John and I gave lectures on all aspects of astronomy, each of which raised several hundred pounds; there were jumble sales, coffee mornings (often at Farthings), postal appeals – you name it, we did it. I presented my first advertisement on television, for Abbey National; they offered me what I regarded as an exorbitant sum, which I immediately handed over. We issued publications, of which one in particular really caught on. For the total solar eclipse of 11 August 1999, visible from Devon and Cornwall, I wrote a booklet, *West Country Eclipse*, which took off like a rocket and raised £25,000 for the Planetarium. Not bad for a couple of days' work.

But we were still desperately short. We had pinned our hopes on the Lottery, because a Planetarium, catering largely for our own youngsters, seemed to be precisely the sort of project which the Lottery ought to support. Not so! We met with flat rejections, because apparently we did not meet any of their required qualifications. The MP for Chichester, Tony Nelson, had been educated at Chichester Boys School and had actually been in the same House with John Mason, so we thought that he might be able to help; he sent us his best wishes.

I did make a suggestion about the Lottery. The organisers are interested mainly in ethnic minorities, feminism and homosexuality, so that if we could work out an appeal incorporating all these features we might stand a chance. Accordingly, I drafted one, pleading the cause of the nomadic Fukawe Indians, in danger of extinction; any extra funds would be diverted to the Planetarium in order to preserve the Indian star lore. Alas, the trustees chickened out, and it was never sent; I wonder what the result would have been? (You may be puzzled at my name for the nomadic Indians, but it shouldn't take you long to work it out!)

I wrote to Tony Blair, who was sympathetic and did put us in touch with a trust which gave us a reasonable sum. All in all the whole saga, from the first idea to the grand opening, spanned seven years. We were determined to have a proper, full-scale Planetarium, ranking with London and Armagh, and we were not prepared to scale it down. The 40-foot dome had to be specially made, and we were able to obtain aircraft tip-back seats; there was a great deal to be done even after the projector had been safely installed.

The projector itself had not been used for some years; we had no circuit diagrams, and the only man who knew about them had emigrated to America and then died. However, John Mason is an expert in these matters, and he coped in a way which was remarkable by any standards. I doubt whether anyone else could have reassembled the projector and put it back into perfect working order.

Outwardly you will not immediately recognize the Planetarium for what it is, because the dome is inside the building – and I assure you that the name of 'The Sir Patrick Moore Building' was not my idea at all; I would have vetoed it, as I was not one of the four most important organisers, but by the time I heard about it, it was a *fait accompli*. Enter through the main door and you will find yourself in an attractive foyer, with various exhibits, a bookstall and also signed photographs of astronauts, Americans, Russians and others, dedicated to the Planetarium. On the upper floor there is a library, a reading room and a computer room. Inside the dome, the projector stays in its recess until it is due to be used, when it is raised and looks very imposing; round the edge of the inner dome there is a representation of the local landscape. The acoustics are excellent; great care was taken about that.

Once the projector was installed, we tested the 'sky', and were delighted; the star images were sharp points, and all the subsidiary projectors, used for details such as meridians, were satisfactory. One or two of the star plates need updating – Sirius, the brightest star in the real sky, is not so prominent in the Planetarium sky as it ought to be, for instance – but we are taking steps to put this right, and even now the effects are virtually as good as they were when the Planetarium was in use at Armagh.

Who should we invite to perform the opening ceremony? There was only one possible choice: Sir Martin Rees, not only one of the great Astronomers Royal, but also a man who has done so much to popularize astronomy. On 5 April 2002 Sir Martin came down from Cambridge, and the South Downs Planetarium was officially launched. There was a large audience, made up of enthusiasts in general plus local dignitaries, and of course those people who had given us help (their names are displayed on plaques inside the Building). Apart from Douglas Denny of the United Kingdom Independence Party, who had been a very strong supporter throughout, local political party leaders were conspicuous by their absence (please remember this at the time of the next General Election!) It was a brilliantly sunny day, and all went well. It was a tribute to the four who had made it all possible: John Mason, John Green, Peter Fray and Roger Prout.

The Planetarium was an instant success, catering for both school parties and the general public. It is not intended to be a commercial venture in the usual sense of the term, but of course it has to cover its costs, and we urgently need new funds to improve the existing facilities and, above all, to recruit paid professional staff. There seems little point in making fresh appeals to the Lottery or to the Ministry of Education or Science; as we are catering for our own people, particularly our own young people, we must 'go it alone'.

It has been worth all the hard work. I wonder how many people will visit us, be encouraged to take up astronomy, and in the end to become leading research scientists? I have a feeling that the answer is: 'A good many'. And even though I was not a key figure, I believe that in being so closely associated with the Planetarium I am preparing to bow out on a high note.

27 Winding Down

On the morning of March the fourth, 2002, I woke up to realize that I had entered my eightieth year. I was far from ecstatic about it, and I remember that when I was a teenager I didn't really believe that anyone could be as old as that. There was also the problem that my wretched physical frame was starting to tell me that it was nearly time to draw stumps.

I always feel that the human body is an unnecessarily complicated piece of mechanism. For example, why do we have to have teeth with nerves? I count myself rather fortunate to have had all mine knocked out when I was eighteen; at least dentures can't ache. I see no point in the appendix, which apparently exists solely to be taken out, and tonsils are simply a nuisance. It would be much better if we developed physically up to the age of, say, twenty-five, and stayed that way until it was time to go, when we would merely vanish with a puff of smoke and a soft sizzle. I can only say that if God designed the human body, he was a rotten mechanic.

I did very well indeed until I was seventy-seven, but then events of long ago caught up with me, and ended all physical games, handling telescopes, playing piano and xylophone, writing properly with a pen, and walking without sticks. It was most annoying, but there was nothing I or anyone else could do about it. Well, I had been given a prediction of no more than ten years of active life; actually I had almost sixty, so I ought not to complain. I can carry on with *The Sky at Night*, and even some lectures; Farthings is never dull, and for the 45th anniversary of *The Sky at Night*, on 29 April 2002, there were over three hundred people there. A good time was had by all, though Jeannie simply curled up on my bed and went to sleep.

Clearly I am on the last lap, but I don't mind; I think I have done most of the things I am capable of doing, and in the words of the famous song I have done things 'my way'. I have left things as tidily as I can, with emphasis on the really important details; notably, I have made sure that whatever happens, Jeannie will always have her tins of Felix.

I have one last trick up my sleeve. I won't have a funeral when I depart from this earthly sphere, my frame can be chopped up and used for research – I will have no further use for it. And in my will, I have left ample funds for my last party to be held at Farthings. I have given instructions for a lighted candle to be placed on the table. A prepared tape will be played, in which I give notice that I am about to do my best to blow out the candle. If I manage it, everyone will know that I am still around.

I made one promise, and I will end by quoting the final words on my tape: 'I won't be seeing you for a bit, but in the long run you will join me. I assure you that as soon as you have rung the bell at the Pearly Gates and signed in, I will be there to meet you. I will at once escort you to the nearest bar, and treat you to a stiff nectar and soda. If you have a tiresome journey I am sure that you will be ready for it, and probably an ambrosia sandwich as well.

'Be seeing you – and all the best!'

'Patrick.'

Index

that I had entered my eightieth year. I was far
c about it, and I remember that when I was a teena
really believe that anyone could be as old as tha
as also the problem that my wretched physical fram
g to tell me that it was nearly time to draw stum
lways feel that the human body is an unnecessaril
ted piece of mechanism. For example why do w
teeth with nerves? I count myself rather fortuna
d all mine knocked out when I was eighteen; at lea
s can't ache. I see no point in the appendix, wh
tly exists solely to be taken out, and tonsils are
a nuisance. It would be much better if we devel
lly up to the age of, say, twenty-five, and stay
y until it was time to go, when we would merely
with a puff of smoke and a soft sizzle. I can on
t if God designed the human body, he was a rotte
c.

id very well indeed until I was seventy-seven, bu
of long ago caught up with me, and ended all phys
handling telescopes, playing piano and xylophone,
properly with a pen, and walking without sticks.
t annoying, but there was nothing I or anyone els